Long-Distance Neutrino Detection–1978

(C.L. Cowan Memorial Symposium, Catholic University)

Clyde L. Cowan

Codiscoverer of the Neutrino
Professor of Physics at Catholic University
Born in Detroit, Michigan, Dec. 6, 1919
Died May 24, 1974

AIP Conference Proceedings
Series Editor: Hugh C. Wolfe
Number 52

Long-Distance Neutrino Detection–1978

(C.L. Cowan Memorial Symposium, Catholic University)

Editors
A.W. Sáenz
Naval Research Laboratory
H. Überall
Catholic University

American Institute of Physics
New York **1979**

L.C. Catalog Card No. 79-52078
ISBN 0-88318-151-7
DOE CONF- 780498

Boys Town Center, Catholic University

Boys Town Center Auditorium

Mrs. Betty Cowan (center); Mrs. Clyde L. Cowan, Sr. (2nd left); son George (left), daughter Marian Kriston and husband (right)

Mrs. Cowan conversing with Georgy Zatsepin (center) and Frederick Reines (right)

Mrs. Clyde L. Cowan, Sr. (left), Mrs. Betty Cowan (center). Background: H. Überall (center), Sam Hanish (right)

A. W. Sáenz

Georgy Zatsepin

Francis J. Kelly

Frederick Reines

(From left) Frederick Reines, Raymond Davis, Jr.

(From left) W. Zachary, H. Überall, S. Hanish

G. Zatsepin (left) conversing with (from left) K. Lande, F. Reines, N. Seeman. Background: L. Wolfenstein, M. Shapiro.

Photos by Marvin T. Jones, Washington DC.

TABLE OF CONTENTS

FOREWORD

F. Reines: Reminiscences: The early days of experimental neutrino physics 1

R. Davis, Jr.: Remembering Clyde Cowan 15

Session I. Solar and Cosmic Neutrinos

Chairman: Carl Werntz, Catholic University

R. Davis, Jr., J. C. Evans, and B. T. Cleveland: The solar neutrino problem 17

Yu. S. Kopysov (reported by G. T. Zatsepin): Solar neutrinos and the catalytic role of a third particle in hydrogen burning 28

M. Deakyne, W. Frati, K. Lande, C. K. Lee, R. I. Steinberg, and E. Fenyves: The Homestake long-range neutrino detector research program 32

D. Eichler: High-energy neutrinos of extraterrestrial origin 38

Session II. Project DUMAND

Chairman: Arthur Roberts, Fermi National Accelerator Laboratory

D. Cline: The study of ultrahigh-energy neutrino interactions in DUMAND 43

J. G. Learned: Project DUMAND and the trade-offs between acoustic and optical detection 58

L. R. Sulak: The acoustic detection of hadronic showers induced by cosmic neutrinos 85

Session III. Neutrino Oscillations, Neutrino Communication, and Contributed Papers

Chairman: M. M. Shapiro, Naval Research Laboratory

A. K. Mann: Neutrino oscillations 101

L. Wolfenstein: Effects of matter on neutrino oscillations 108

F. J. Kelly, A. W. Sáenz, and H. Überall: Telecommunication by high-energy neutrino beams 113

R. Ehrlich: Time dependence of the Brookhaven solar neutrino counting rate and the neutrino-oscillation hypothesis 129

D. Leiter: Fermion nonminimal gravitational coupling and the solar neutrino problem 139

FOREWORD

A conference on the subject of long-distance neutrino detection, co-sponsored by The Catholic University of America and the Naval Research Laboratory, both at Washington, DC, was held in the auditorium of the new Boys Town Center on the Catholic University Campus on April 25, 1978. Sessions were held on the present experimental status of the detection of solar and cosmic-ray neutrinos, on the proposed DUMAND project for detecting ultrahigh energy neutrinos of cosmic origin in the ocean, on experimental and theoretical aspects of neutrino oscillations, and on a proposed method of telecommunication by high-energy neutrino beams.

The meeting was dedicated to the memory of the late Clyde L. Cowan, co-discoverer (with Fred Reines) of the neutrino, who had been a Professor of Physics at Catholic University from 1959 until his untimely death in 1974. While at this institution, Clyde Cowan had remained interested in neutrino physics, and in fact had been the first to suggest (in 1965) the possibility of detecting high-energy neutrinos in the ocean by means of the Čerenkov light given off by their charged reaction products, a suggestion which is basic to many of the detection schemes proposed for both the DUMAND and the neutrino communications projects, and whose feasibility has been proved in the 1967 prototype experiment of Gordon K. Riel of the Naval Surface Weapons Center, White Oak, MD. Besides his scientific accomplishments (of which the discovery of the neutrino alone would have well merited him the Nobel award), Clyde Cowan also showed deep human qualities, such as his fatherly concern for the progress and the well-being of his students, and his constant helpfulness towards his colleagues, collaborators, and scientific employees. Those of us who had the privilege of knowing him will remember him for these qualities just as much as for his stature and his enthusiasm in the scientific domain. In fact, the two papers at the beginning of this volume (by Fred Reines and by Raymond Davis, Jr.), which reminisce on their personal relationships with Clyde during the period of the discovery of the neutrino, also serve well in presenting a vivid glimpse of Clyde Cowan as a person.

To return to the subject matter of the conference, it was intended here to demonstrate the far-ranging nature of neutrino beams as a means of "communication" in the widest sense. They can reach us, undeflected, from the farthest corners of the universe and, in contrast to light rays, they can also penetrate immense amounts of matter on their way. Hence, they can give us information on what is happening at the center of the sun or of distant stars. They can tell us about supernovae, nebulae, and galaxies. And besides opening up for us a whole new discipline, Neutrino Astronomy, they may serve us in a more mundane sense, yet on a global scale, by carrying messages straight through the earth to otherwise inaccessible places.

We wish to thank all those who gave invited and contributed talks for their willingness to share with us the information they obtained in their most recent studies on the mentioned subjects; and we thank Professor Carl W. Werntz (Catholic University), Dr. Arthur Roberts (Fermi National Accelerator Laboratory), and Dr. Maurice M. Shapiro (Naval Research Laboratory) for their willingness to chair the three sessions of the symposium. The hospitality of the Boys Town Center of Catholic University and of its director, Professor James P. O'Connor, as well as of the Catholic University Physics Department is appreciated. We are grateful to our Naval Research Laboratory colleagues Dr. Robert H. Bassel, Dr. Francis J. Kelly, Dr. Maurice M. Shapiro, Dr. Rein Silberberg, and

Dr. W. W. Zachary for valuable comments, pertinent to the editing of these proceedings. The efforts of Miss E. Pickenpaugh and Mr. R. M. Harmon in identifying several ambiguous references is also appreciated. Last, but not least, we wish to thank the Naval Research Laboratory for providing financial and other support for the symposium and to the Office of Naval Research for their active interest.

A. W. Sáenz

H. Überall

REMINISCENCES: THE EARLY DAYS OF EXPERIMENTAL NEUTRINO PHYSICS*

Frederick Reines
Department of Physics, University of California, Irvine, California 92717

It is now more than 45 years since Pauli[1] (1930) and Fermi[2] (1933) formulated the neutrino hypothesis and 25 years since Clyde Cowan and I and our colleagues at Los Alamos[3-5] made the first, tentative observation of the free neutrino at a Hanford reactor via the inverse beta process

$$\nu + p \rightarrow n + e^{+}.$$

Our choice of this reaction was most felicitous because of its simplicity, distinctive products, and the scintillation properties of some liquid hydrocarbons.[6] Three years later (1956), we completed the job in a definitive way at the Savannah River Plant and experimental neutrino physics was launched.[9]

In the summer of 1951, after several months of concentrated thought, I decided that the detection of the elusive neutrino was a goal worth striving for. (At the time, the neutrino hypothesis was already firmly fixed in the lexicon of physics. Physicists generally believed that its existence was demonstrated, albeit indirectly, and that, in fact, it is not directly observable. It was argued that the neutrino hypothesis explained the apparent lack of energy-momentum conservation in beta decay, e.g., shapes of decay spectra and nuclear recoil in K capture, and hence that the neutrino existed. Had any measurements made on the beta-decay process been found to be inconsistent with the ν hypothesis, then it could have been argued that the ν does not exist. The converse is untrue. In fact, reasoning which involves only observations at the "scene of the crime" is circular or, at best, incomplete. However attractive the neutrino hypothesis is as an "explanation" of beta decay, the proof of existence must be derived from an observation made at a location other than that at which the decay process occurs (the neutrino must be observed in its free state to induce inverse beta decay or otherwise interact with matter at a remote point).

A few rough estimates indicated to me that a nuclear explosion, with its intense pulse hopefully overriding backgrounds, was the most promising source and that a suitably shielded detector of mass ~ 1 ton might do the job. In any event, it would be a vastly more sensitive search than previously imagined. I had no idea how such an incredibly large detector could be made and thought it might be helpful to talk with Fermi, who was spending the summer at Los Alamos. As it turned out, he did not have any idea, either, and that ended the matter — almost.

Some months later, Clyde and I, who had met while working on nuclear weapons tests (Operation Greenhouse, 1951), found ourselves grounded in Kansas City while on our way to Princeton to learn from Lyman Spitzer the status of the program there on controlled thermonuclear reactions. During the long wait for aircraft repairs, we discussed various problems on which it would be nice to work, and I mentioned my thoughts on the neutrino. His reaction was immediate — there must be a way to make such a detector! Our partnership began at that point, and our ideas flowed together in a mutually reinforcing manner which often made it difficult to decide who thought of what.[11] I recall one instance which illustrates the depth of our collaboration and, incidentally, supports the view that the observation of the neutrino might have been delayed

*Supported in part by the United States Department of Energy.

ISSN:0094-243X/79/520001-14$1.50

but for our work. We gave a talk to the Physics Division at Los Alamos in which we described our ideas for the use of an as yet nonexistent large liquid scintillator[12] — nicknamed "El Monstruo" — in the vicinity of a nuclear explosion. We mentioned the use of the delayed coincidence between the positron and neutron pulses as a label for the reaction — it had not yet occurred to us that the label could be used to reduce the background. Dr. J. M. B. Kellogg, the Physics Division leader, asked whether it might not be possible to use a fission reactor instead of a bomb. We argued that it was not — and, besides, Fermi and Bethe had agreed with us that the bomb was a most promising source!

That night I phoned Clyde and we told each other how the delayed coincidence could be used to reduce the background making the reactor an attractive source. We immediately altered our plans and the next morning met with Kellogg to cancel our bomb preparations and arrange to develop and build a detector suitable for the Hanford reactor. We learned later that others attending the talk had considered Kellogg's question and concluded that the bomb was better suited than the reactor. You can well imagine how embarrassing it would have been had the roles been reversed!

A letter to Fermi telling him of our reactor proposal elicited the response shown in Fig. 1.

Viewed from the perspective of today's computer-controlled kiloton detectors, NaI-crystal palaces, giant accelerators, and 50-man groups our efforts to detect the neutrino appear quite modest. In the 1950s, however, our work was thought to be large-scale. The idea of using 90 photomultiplier tubes and detectors large enough to enclose a human was considered to be most unusual. We faced a host of unanswered questions. Was the scintillator sufficiently transparent to transmit its light for the necessary few meters? How reflective was the paint? How could one add a neutron capturer without poisoning the scintillator? Would the tube noise and afterpulses from such a vast number of P.M. tubes mask the signal? etc., etc. And besides, were we not monopolizing the market on P.M. tubes? (As it turned out, we were not — headlight dimmers on Cadillacs consumed far more P.M. tubes than we did!). In the search for answers to these questions, we received strong support from various scientists at Los Alamos and Clyde made good use of his undergraduate training as a chemist and his considerable abilities with electronics.

It soon became clear that this new detector, designed solely for neutrinos, had unusual properties with regard to other particles as well, e.g., neutron and gamma-ray detection efficiencies near 100%. We recognized that such a detector could be used to study such diverse quantities as neutron multiplicities in fission, muon capture, muon-decay lifetimes, and the natural radioactivity of human beings. We measured the activity of some humans, pointed out other uses and continued with our neutrino search. (These other applications have since been made by other workers.)

Our entourage arrived at Hanford in the spring of 1953 (Figs. 2,3,4,5). After a few months of operating, during which we made several restackings of hundreds of tons of specially fabricated boron-paraffin boxes and lead-brick shielding, we concluded that we had done all we could in the face of an enormous reactor-independent background. Bone-tired, we turned off the equipment and took the train back to Los Alamos knowing that we had done our best, but not knowing that we had actually measured a hint of a signal! (Table I).

THE UNIVERSITY OF CHICAGO
CHICAGO 37 · ILLINOIS
INSTITUTE FOR NUCLEAR STUDIES

October 8, 1952

Dr. Fred Reines
Los Alamos Scientific Laboratory
P.O. Box 1663
Los Alamos, New Mexico

Dear Fred:

Thank you for your letter of October 4th by Clyde Cowan and yourself. I was very much interested in your new plan for the detection of the neutrino. Certainly your new method should be much simpler to carry out and have the great advantage that the measurement can be repeated any number of times. I shall be very interested in seeing how your 10 cubic foot scintillation counter is going to work, but I do not know of any reason why it should not.

Good luck.

Sincerely yours,

Enrico

Enrico Fermi

EF:vr

Fig. 1. Letter from Fermi on hearing about our plan to use the reactor.

Hanford
(1953)

Fig. 2. First large (0.3 m^3) liquid scintillation detector in shield. The liquid was viewed by ninety 2'' photomultiplier tubes. Prior to the development of this detector a 0.02 m^3 volume was considered "large".

Fig. 3. Shield configuration. The members of the group for the Hanford phase of the search are listed on the "Project Poltergeist" sign.

Hanford
(1953)

Fig. 4. Clyde L. Cowan, Jr., making a connection in the back of the shield.

Fig. 5. F. Reines (left) and C. L. Cowan at the control console.

Table I. Listing of data, Hanford Experiment.

Run	Pile status	Length of run (sec)	Net delayed pair rate (counts/min)	Accidental background rate (counts/min)
1	up	4000	2.56	0.84
2	up	2000	2.46	3.54
3	up	4000	2.58	3.11
4	down	3000	2.20	0.45
5	down	2000	2.02	0.15
6	down	1000	2.19	0.13

Reactor up (10,000 seconds): 2.55 ± 0.15 counts/min.
Reactor down (6,000 seconds): 2.14 ± 0.13 counts/min.
Difference = 0.41 ± 0.20 counts/min.

Back home, we puzzled as to the origin of the reactor-independent signal. Was it due to natural neutrinos? Could it be due to fast neutrons from the nuclear capture of cosmic ray muons? The easiest way to find out was to put the detector underground and we did, showing it to be from cosmic rays. While we were engaged in this background test some theorists were constructing a world made predominantly of neutrinos!

Encouraged by the tentative result at Hanford, we regrouped and designed and constructed a detector which would employ the detailed characteristics of the $\overline{\nu}_e$ + p reaction and so discriminate more selectively against backgrounds, both reactor independent and reactor-associated. The detector was completed in 1955 and, at the suggestion of John A. Wheeler, taken to a newly completed reactor at the Savannah River Plant, where a definitive observation of $\overline{\nu}_e$ + p was made[9] (1956) (Figs. 6,7,8,9,10).

I remember the first turn-on of the detector at SRP. No signals were seen! It was a most peculiar feeling. Maybe there were no neutrinos! Maybe they existed but were unstable and didn't reach our detector. We continued tuning the detector and the signal appeared, but what a heady — if unwarranted — flight of fancy.

After convincing ourselves by a redundant series of tests[17] that we were observing the neutrino, we decided to let Pauli know how correct he was (Fig. 11). Shortly after detection of the $\overline{\nu}_e$, evidence was found by Wu et al., Garwin et al., and Friedman and Telegdi[18] for parity non-conservation in beta decay (1957). This was explained by Lee and Yang [19,20], Landau, [21] and Salam[22] as associated with the two component character of the neutrino, which, incidentally, predicts an increase by a factor of two in the $\overline{\nu}_e$ + p cross section over that of the parity-conserving, four-component neutrino. It is amusing to speculate on the credibility of this explanation for the violation of parity had it been put forward prior to our proof of the ν's existence. It is also interesting to reflect on the

Savannah
River
(1955-6)

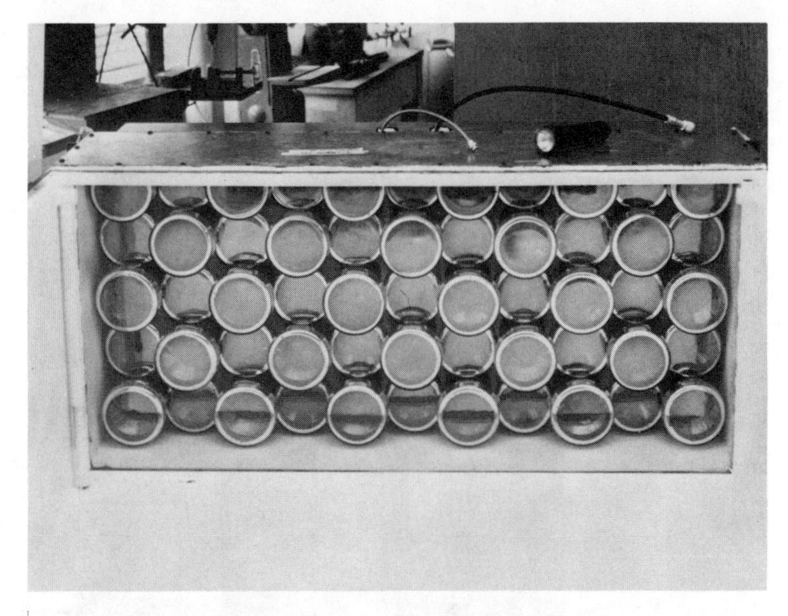

Fig. 6. Inside view of one of the three detector tanks used in neutrino experiments at Savannah River Plant. The tank had fifty-five 5'' photomultiplier tubes at each end.

Fig. 7. Portable scintillation liquid storage tanks, 3600 gallon capacity.

Savannah River (1955-6)

Fig. 8. Inside view of electronics van showing equipment required to select and record neutrino signals.

Fig. 9. End view of neutrino detector array inside massive lead shield.

Savannah
River
(1955-6)

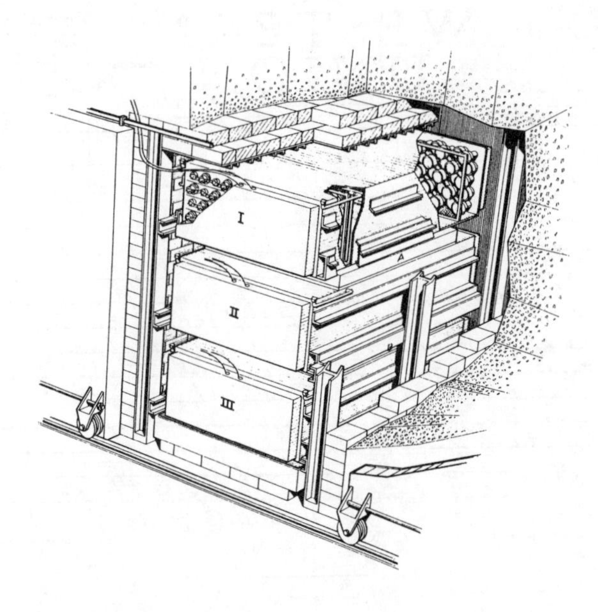

Fig. 10. Sketch of detectors inside their lead shield. The tanks marked I, II, and III contained 1400 liters of triethylbenzine (TEB) liquid-scintillator solution which was viewed in each tank by 110 five-inch photomultiplier tubes. The TEB was made to scintillate by the addition of p-terphenyl (3g/ℓ) and POPOP wavelength shifter (0.2 g/ℓ). The tubes were immersed in pure nonscintillating TEB to make light collection more uniform. Tanks A and B were polystyrene and contained 200 ℓ of water which provided the target protons and in which were dissolved as much as 40 kg of $CdCl_2$ to capture the product neutrons.

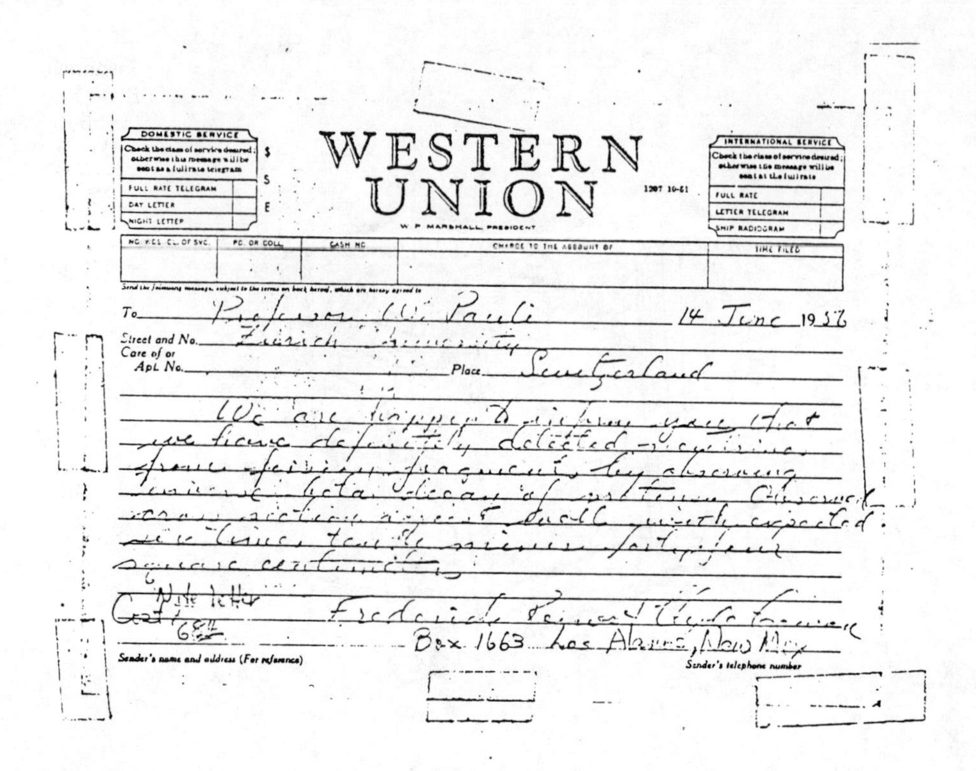

DOMESTIC SERVICE
Check the class of service desired; otherwise this message will be sent as a full rate telegram
FULL RATE TELEGRAM
DAY LETTER
NIGHT LETTER

WESTERN UNION

W. P. MARSHALL, PRESIDENT

1207 10-51

INTERNATIONAL SERVICE
Check the class of service desired; otherwise the message will be sent at the full rate
FULL RATE
LETTER TELEGRAM
SHIP RADIOGRAM

NO. WDS.-CL. OF SVC.	PD. OR COLL.	CASH NO.	CHARGE TO THE ACCOUNT OF	TIME FILED

Send the following message, subject to the terms on back hereof, which are hereby agreed to

To Professor W. Pauli 14 June 1956
Street and No. Zurich University
Care of or Apt. No. Place Switzerland

We are happy to inform you that we have definitely detected neutrinos from fission fragments by observing inverse beta decay of protons. Observed cross section agrees well with expected six times ten to minus forty-four square centimeters.

Night letter
Cost 68¢
Frederick Reines Clyde Cowan
Box 1663 Los Alamos, New Mex

Sender's name and address (For reference)
Sender's telephone number

Fig. 11. Telegram to Pauli informing him of our results. Typed text of telegram is given on the following page.

Text to Fig. 11

Telegram to Pauli
sent by Western Union

To: Professor W. Pauli
Zurich University, Switzerland

June 14, 1956

We are happy to inform you that we have definitely detected neutrinos from fission fragments by observing inverse beta decay of protons. Observed cross section agrees well with expected six times ten to minus forty four square centimeters.

Frederick Reines and Clyde Cowan
Box 1663 Los Alamos, New Mexico

Nite letter
Cost $6.84

*Note: Pauli was at the Eidgenössische Technische Hochschule, not Zurich University, but the message was forwarded to him at CERN, where he interrupted the meeting he was attending by reading the telegram to the conferees and then made some impromptu remarks regarding the discovery.

We received no reply from Pauli but learned later that a case of champagne was consumed in celebration by him and some friends.

Two years later, on hearing we were in Geneva, Pauli invited us to visit him in Zurich. Clyde was unable to attend, but I spent a most pleasant hour with Pauli. His gentle manner on that occasion belied the stories about his acid and difficult nature.

fact that evidence for the parity factor would have been obtained in due course,[23] independently of the $\theta - \tau$ puzzle which led Lee and Yang to question the conservation of parity in the weak interaction.

Once detected, it occurred to us that the free neutrino could be used to probe the weak interaction in energy ranges inaccessible to ordinary beta decay if sufficiently potent accelerators could be constructed. Also, we puzzled, why should the "neutretto" of $\pi \rightarrow \mu$ decay be the same as the "neutrino" of nuclear beta decay — Occam's razor notwithstanding? We suggested (1957) that it might be nice to go to the Brookhaven accelerator with a suitable detector but were unsuccessful in persuading the authorities at Los Alamos to let us continue our search.[25]

At this juncture, Clyde decided to leave Los Alamos and our exciting years together came to an end.

Clyde Cowan left a legacy which will live in the annals of physics. In the search for the neutrino, he exhibited the courage to tackle a problem which had defied solution for 20 years and the creative imagination to contribute in a fundamental way to its solution and the founding of an exciting and fruitful field of physics.

It was a privilege to be his colleague in those days and his friend for the remainder of his life.

FOOTNOTES AND REFERENCES

1. W. Pauli Jr., address to the Group on Radioactivity, Tübingen, December 4, 1930 (unpublished); Rapports du septième conseil physique Solvay, Bruxelles, 1933 (Gauthier-Villars, Paris, 1934), Chap. 1, p. 324.
2. E. Fermi, Z. Phys. 88, 161 (1934).
3. F. Reines and C. L. Cowan, Jr., "A Proposed Experiment to Detect the Free Neutrino," Phys. Rev. 90, 492 (1953).
4. C. L. Cowan Jr., F. Reines, F. B. Harrison, E. C. Anderson and F. N. Hayes, "Large Liquid Scintillation Detectors," Phys. Rev. 90, 493 (1953).
5. F. Reines and C. L. Cowan, Jr., "Detection of the Free Neutrino," Phys. Rev. 92, 830 (1953).
6. B. Pontecorvo[7] (1946) and L. W. Alvarez[8] (1949) suggested a radio-chemical method using a fission reactor based on the reaction $\nu + {}^{37}\mathrm{Cl} \rightarrow {}^{37}\mathrm{Ar} + e^-$. The method was not pursued by them. In the case of Alvarez, he was dissuaded by his estimates of the background anticipated from cosmic rays, estimates which later proved to be correct. As we know now, the ν produced in fission is $\bar{\nu}_e$, and the ν required for the ^{37}Cl reaction is ν_e, so that the reactor result would have been negative even though the neutrino exists. Incidentally, the subsequent solar-neutrino search of Raymond Davis, Jr., using this reaction was made logically possible by the demonstration that the neutrino exists.
7. B. Pontecorvo, Inverse Beta Decay (Division of Atomic Energy, National Research Council of Canada, Chalk River; declassified and issued by the Atomic Energy Commission in 1949). This Canadian report was based on a lecture of November 30, 1946.
8. L. W. Alvarez, University of California Radiation Laboratory Report No. UCLR-328, 1949 (unpublished).
9. C. L. Cowan Jr., F. Reines, F. B. Harrison, H. W. Kruse, A. D. McGuire, "Detection of the Free Neutrino: A confirmation," Science 124, 103 (1956). A full paper appeared in Phys. Rev. 117, 159 (1960).
10. It is, of course, not possible to know how the field would have developed if Clyde and I had not met and decided to work together on this "manifestly impossible" search, but in view of the general absence of activity in this direction at the time I suspect that the observation would have been somewhat delayed.
11. However, I do remember one conversation regarding detection techniques most vividly. "Why not, suggested Clyde, "make a device analogous to a cloud chamber, but of liquid, to obtain the necessary target mass, and use it in our search." We discussed it at some length, but discarded it because it could not be triggered by the event and random triggers would give a small duty cycle. It was a nice idea, as subsequent events have demonstrated, but it did not suit our purpose, which was to detect the neutrino, and we did not pursue it. (As we now know, the bubble chamber was invented about that time by Donald Glaser, and in the hands of Alvarez and others turned out to be extremely useful for particle physics, eventually including neutrinos at accelerators.)
12. The technique of scintillation counting followed the discovery (1903) by W. Crookes, and by J. Elster and H. Geitel,[13] of the scintillation properties of zinc sulphide exposed to alpha particles (described in E. Rutherford, J. Chadwick and C. D. Ellis, Radiations from Radioactive Substances (Cambridge University Press, 1930)). It received great impetus from the development of the photomultiplier tube and the crucial observation by Kallmann[14] and others[15,16] (1950) that liquids could be made to scintillate with high efficiency when the scintillating compound

was at low concentration. Our contribution was to recognize that given a sufficiently transparent scintillator and enough photocathode area, one should, in principle, be able to make a detector of almost arbitrarily great size—just what was needed for neutrino detection.

13. J. Elster and H. Geitel, Phys. Z. 4, 439 (1903).
14. H. Kallmann, Phys. Rev. 78, 621 (1950).
15. M. Ageno, M. Chiozotto, and R. Querzoli, Atti Accad. Naz. Lincei Cl. Sci. Fis. Mat. Nat. Rend. 6, 626 (1949); Phys. Rev. 79, 720 (1950).
16. G. T. Reynolds, F. B. Harrison, and G. Salvini, Phys. Rev. 78, 488 (1950).
17. The observed reactor-associated rate of pulse pairs was tested as arising from neutrino interactions by demonstrating that the first pulse was due to the annihilation of a positron by an electron, the second pulse was due to the capture of a neutron, the signal was proportional to the number of protons in the water target, and the signal was not due to gamma rays and neutrons leaking through the reactor shield.
18. C. S. Wu, E. Ambler, R. W. Hayward, D. D. Hoppes, and R. F. Hudson, Phys. Rev. 105, 1413 (1957).
19. T. D. Lee and C. N. Yang, Phys. Rev. 104, 254 (1956).
20. T. D. Lee and C. N. Yang, Phys. Rev. 105, 1671 (1957).
21. L. Landau, Nucl. Phys. 3, 127 (1957).
22. A. Salam, Nuovo Cimento 5, 299 (1957).
23. We measured the cross section[24] in the Fall of 1956 with equipment designed and built for the purpose in 1954-5, but did not publish the result until a measurement had been made of the $\bar{\nu}_e$ spectrum from fission (1957), thus enabling a comparison to be made with theory.
24. F. Reines and C. L. Cowan, Jr., "Measurement of Free Antineutrino Cross Section"; R. E. Carter, F. Reines, J. J. Wagner, and M. E. Wyman, "Expected Cross Section from Measurements of Fission Fragment Electron Spectra." These constitute Parts I and II of a paper on "The Free Antineutrino Absorption Cross Section," in Second United Nations Conference on the Peaceful Uses of Atomic Energy (United Nations, New York, 1958), A/Conf. 15/P/1026. See also Phys. Rev. 113, 273 (1959); 113, 280 (1959).
25. The idea to use an accelerator was independently conceived by B. Pontecorvo[26] (1959) and by M. Schwartz[27] (1960). It was demonstrated[28] (1962) at Brookhaven by Schwartz and his colleagues that $\nu_\mu \neq \nu_e$. Groups at CERN[29,30] (1964), using a meson-focussing magnet, obtained a neutrino beam of 100 times higher intensity. They detected not only the muonic neutrino ($\nu_\mu, \bar{\nu}_\mu$), but also electron production by the small admixture in the ν beam stemming from β-decays of K-mesons, thus verifying the existence of ν_e.
26. B. Pontecorvo, "Electron and Muon Neutrinos," reprint p-376 of JINR, Dubna, U.S.S.R. (1959); Zh. Eksper. Teor. Fiz. 6, 1751 (1959).
27. M. Schwartz, Phys. Rev. Lett. 4, 306 (1960).
28. G. Danby, J. M. Gaillard, K. Goulianos, L. M. Lederman, N. Mistry, M. Schwartz, and J. Steinberger, Phys. Rev. Lett. 9, 36 (1962).
29. M. M. Block, H. Burmeister, D. C. Cundy, B. Eiben, C. Franzinetti, J. Keren, R. Mollerud, G. Myatt, M. Nikolic, A. Orkin Lecourtois, M. Paty, D. H. Perkins, C. A. Ramm, K. Schultze, H. Sletten, K. Soop, R. Stump, W. Venus, and H. Yoshiki, Phys. Lett. 12, 281 (1964).
30. J. K. Bienlein, A. Bohm, G. Von Dardel, H. Faissner, F. Ferrero, J. M. Gaillard, H. J. Gerber, B. Holm, V. Kaftanov, F. Krienen, M. Reinharz, R. A. Salmeron, P. G. Seiler, A. Staude, J. Stein, and H. J. Steiner, Phys. Lett. 13, 80 (1964).

REMEMBERING CLYDE COWAN

Raymond Davis, Jr.
Brookhaven National Laboratory, Upton, New York 11973

I am very pleased to be invited to this symposium on neutrino physics dedicated to Clyde Cowan. I remember the first occasion that I met Clyde and Fred. It was when they presented their ideas and initial Hanford results at the American Physical Society meeting at Columbia University in 1954. It is interesting that their talk was listed as a 30 minute talk to be shared by Clyde and Fred. A year or so after that, Clyde and Fred were setting up and carrying out their famous experiment at the Savannah River plant. There are two reactors at Savannah River that are almost ideally suitable for neutrino experiments. They planned to use the P-reactor and I was setting up an experiment with 1000 gallons of carbon tetrachloride at the R-reactor. I remembered seeing the apparatus Fred described to you when it arrived. The pictures Fred showed[1] bring back memories of over 20 years ago. On my periodic visits I would go into their trailer, the one Fred showed that was packed solid with electronics. We would spend a couple of hours discussing progress on our respective experiments. On one occasion, they invited me over to their reactor, closed the door, made me promise not to tell anyone, and then told me for the first time that they were convinced that they had a positive signal of approximately the correct magnitude. It was clear by then that my experiment would not detect antineutrinos, but it did serve as a check on the principle of lepton conservation. Their experiment was beautifully designed to observe delayed coincidence events and I have always believed that this Savannah River experiment was the first one to definitely observe neutrino interactions. It was indeed a rare privilege to be at Savannah River during this period and witness the development of one of the first large scale particle physics experiments. Their work was the first to apply very large liquid scintillators in particle and cosmic-ray physics experiments. Also of perhaps greater importance, their experiment was a novel application of the delayed-coincidence technique that has been of enormous value.

After their Savannah River experiment, their active collaboration ended when they left Los Alamos, but they were still closely associated in spirit and attitudes on new ideas. I visited Clyde many times at Catholic University and met him numerous times at scientific meetings. Clyde was always interested in the work of other scientists, and talking to him was lively, easy, and pleasant. I have felt closely related to Clyde and Fred in our mutual interest in neutrino experiments, and our long standing friendship. Clyde was a scientist with imagination, interested in new and exciting ideas in science. He developed an intense interest in astrophysics and carried out one of the early experiments searching for neutrinos from space. Like many first attempts in this impossibly difficult field, it did not observe neutrinos (or neutrals) from space. However, it was a beautiful experiment and, again, was a first in experimental technique. Science is greatly dependent upon people like Clyde Cowan who have intense interest and originality, and are willing to devote themselves to speculative new ideas.

I have been asked to give a review of the solar neutrino problem at two conferences within a week. The paper[2] that my colleagues and I will present at this symposium will be identical to the one presented at the Neutrinos—78 Conference at Purdue University.

REFERENCES

1. F. Reines, preceding paper.
2. R. Davis et. al., following paper.

ISSN:0094-243X/79/520015-01$1.50

THE SOLAR NEUTRINO PROBLEM*

R. Davis, Jr., J. C. Evans, and B. T. Cleveland
Brookhaven National Laboratory, Upton, New York 11973

ABSTRACT

A summary of the results of the Brookhaven solar neutrino experiment is given and discussed in relation to solar-model calculations. A review is given of the merits of various new solar neutrino detectors that have been proposed.

INTRODUCTION

We would like to review the present status of the solar neutrino problem. First will be a report on the Brookhaven ^{37}Cl detector that has been in operation for 10 years. The results obtained during the last 7 years will be compared with the current solar-model calculations. In recent years, a number of new solar-neutrino detectors have been proposed. These various detectors will be discussed in light of some of the current ideas on solar models and neutrino properties.

The sun is generating energy principally by the proton-proton chain of reactions. The neutrinos are produced by the P-P reaction and a few beta decay processes. These reactions and decay processes are listed in Table I along with the neutron energy spectra and fluxes at the earth. The highest flux (6×10^{10} cm^{-2} sec^{-1}) arises from the P-P reaction, but these neutrinos have very low energies (< 0.4 MeV). There are a group of processes emitting neutrinos with energies up to 1.7 MeV, with fluxes in the range of 2-34 $\times 10^8$ cm^{-2} sec^{-1}. The neutrinos from ^8B decay have relatively high energies, but the flux of these neutrinos is very low (3×10^6 cm^{-2} sec^{-1}). Since the solar neutrinos have very low energies and their flux at the earth is low, the only means of observing them that has been developed is a radiochemical technique based upon the inverse-beta processes. We developed a radiochemical detector based upon the neutrino-capture reaction $^{37}Cl(\nu,e^-)^{37}Ar$. A large detector, capable of observing the calculated solar neutrino flux, was built in the period 1964-1967. Actually, at the time three direct counting detectors were built. Two were based on inverse-beta processes, $D(\nu,e^-H)H$ and $^7Li(\nu,e^-)^7Be$, respectively, and one was based upon neutrino-electron scattering.[1] It is interesting to note that these experimental approaches to observing the energetic ^8B solar neutrinos are now being reconsidered. We would like now to give you a summary of the present results from the ^{37}Cl experiment.

THE ^{37}Cl EXPERIMENT

Table I shows the flux-cross section product for each of the neutrino sources in the sun. The fluxes are those derived from the standard solar model,[2] and the cross sections are from Bahcall.[3] The total neutrino capture rate expected from standard solar model calculation is 4.7 SNU, where SNU represents a solar neutrino unit (SNU $\equiv 10^{-36}$ captures/sec- ^{37}Cl atom). The Brookhaven detector contains 615 tons of liquid C_2Cl_4 or 2.18×10^{30} atoms of ^{37}Cl. The expected solar neutrino capture rate is 0.88 per day from the current standard-solar-model calculations. The detector is located deep under-

*Research performed at Brookhaven National Laboratory under contract with the U.S. Department of Energy and supported by its Division of Basic Energy Sciences.

ISSN:0094-243X/79/520017-11$1.50

ground to reduce the production of ^{37}Ar in the liquid from cosmic-ray muons. It is located in the Homestake Gold Mine at Lead, SD, at a depth of 4850 feet, corresponding to 4400 hg/cm^2 of overhead shielding. The tank containing the liquid is also shielded with water to eliminate the production of ^{37}Ar from fast neutrons from the surrounding rock wall.

Table I. Solar neutrino fluxes[2] and cross sections[3] for $^{37}Cl(\nu,e^-)^{37}Ar$.

Neutrino sources and energies in MeV	Flux on earth ϕ in cm^{-2} sec^{-1}	Cross section σ in cm^2	Capture rate in ^{37}Cl, $\phi\sigma \times 10^{36}$ sec^{-1} SNU
$H+H \rightarrow D+e^+ + \nu$ (0-0.42)	6.1×10^{10}	0	0
$H+H+e^- \rightarrow D+\nu$ (1.44)	1.5×10^8	1.54×10^{-45}	0.23
^7Be decay (0.86)	3.4×10^9	2.4×10^{-46}	0.80
^8B decay (0-14)	3.2×10^6	1.08×10^{-42}	3.46
^{15}O decay (0-1.74)	1.8×10^8	6.6×10^{-46}	0.12
^{13}N decay (0-1.19)	2.6×10^8	1.6×10^{-46}	0.04
			$\Sigma\phi\sigma$ = 4.65

The ^{37}Ar is removed from the tank periodically by purging with helium gas. Argon is collected from the helium stream by a charcoal filter. It is finally purified by gas chromatography, gettered with hot titanium, and placed in a small low-level proportional counter to observe the ^{37}Ar-decay events. The detailed procedures are described in earlier reports.[4] The small proportional counter (internal volume 0.6 cm^3) is operated in anticoincidence with a well-type sodium iodide scintillation counter to eliminate cosmic-ray events. The counters are operated inside a 20 cm thick mercury shield. Pulse rise-time, pulse height, and time of occurrence are recorded for each count, along with auxiliary information to check the performance of the recording system. ^{37}Ar-decay events produce a fast-rising pulse that can be clearly distinguished from background events from beta rays and Compton electrons. The ^{37}Ar-like events are thereby characterized by their energy and pulse rise-time. Individual samples are counted for long periods of time, usually 150 to 250 days, so that the decay of ^{37}Ar (half-life 35 days) can be observed. During the entire period, only a small number of counts are recorded (12 on the average). The time of occurrence of the counts with the characteristic energy and rise-time was treated by a maximum likelihood statistical treatment developed by one of us (B.T.C.) to separate the 35 day decaying component from the presumed constant background counting rate. This treatment yields a most likely value for the ^{37}Ar production rate in the tank and includes fluctuations (1) in the ^{37}Ar production in the tank, (2) in the decay during the production period, (3) during the extraction, and (4) in the counting. The errors were obtained by taking the upper and lower bounds defined by 34% of the total area under the likelihood function on either side of the most likely value. In the event that the most likely value is too low for this procedure to be followed, the upper error given corresponds to the bound that includes 68% of the area under the likelihood function. To obtain an average of a number of runs, one uses the likelihood function formed by multiplying the separate likelihood functions for each run.

The ^{37}Ar-production rates derived from this analysis are shown in Fig. 1. These are 30 individual runs, Nos. 18-51, that were made from 1970 to 1977. Prior to run 18, we did not use pulse rise-time discrimination; results from these earlier experiments are given in Ref. 5. Every long exposure run is given except run 23; it was a poor run due to a valve leak. The missing run numbers correspond to runs in which special tests were performed. The average ^{37}Ar production for all runs shown is 0.41 ± 0.06 ^{37}Ar atoms per day in 615 tons of C_2Cl_4. There is a cosmic-ray background production of ^{37}Ar in the tank from muons and cosmic-ray produced muon neutrinos that must be subtracted to obtain the ^{37}Ar production rate that we could assign to solar neutrinos.[6] The results are as follows:

		^{37}Ar atoms/day
Average ^{37}Ar production rate (18-51)	=	0.41 ± 0.06
Cosmic ray background (muons and ν_μ)	=	0.08 ± 0.03
Rate above known backgrounds		0.33 ± 0.03

Possible solar neutrino rate =
$5.31 \times (0.33 \pm 0.07) = 1.75 \pm 0.4$ SNU

It is interesting to see if there is any change in the neutrino flux during the last 7 years that is observable with the ^{37}Cl experiment. Theoretically, there is no reason to expect any change in the solar neutrino flux on time scales less than about 10^4 years. We have made yearly averages and these are presented in Fig. 2. Run 27, yielding our highest experimental point, was a long exposure and, therefore, dominates the 1972 value. Shown for 1972 are two values, one including run 27 (dotted), and one without it. From this plot it is evident that there has not been any change in the ^{37}Ar production rate outside of our statistical errors.

COMPARISON WITH THE STANDARD SOLAR MODEL

One can compare this result with the generally accepted solar-model calculation of 4.7 SNU. It is difficult to assign an error to the standard-solar-model calculation. If the errors in the various input data used in the calculation are evaluated, one can estimate an error of about ±30%.[7] The standard solar model presumes that the sun is a spherical nonrotating body with an initial composition identical to that observed now in its photosphere. The structure is derived from a set of differential equations for hydrostatic equilibrium, for radiation transport, and for energy production. It is assumed that the energy is derived solely from the thermal fusion reactions of the P-P and CNO chains. An ideal equation of state is used and the kinetic velocities are considered to be accurately Maxwellian. Various data are used in these calculations. Some are very well determined, such as the mass, age, and luminosity of the sun, and others are not as well known. The laboratory-derived nuclear reaction cross sections are used, and the theoretically calculated opacities. The standard model predicts that the sun is operating on the P-P cycle and less than 2 percent of the energy is produced by the CNO cycle. This prediction agrees with our experiment, since if the sun were operating on the CNO cycle the neutrino capture rate would be 25 SNU. Another conclusion is that the luminosity of the sun increases with time at the rate of 5% per billion years. This result has been discussed in relation to the earth's climate,[8] and some have concluded that, if the earth's atmospheric composition has not changed, a 5% drop in solar luminosity would cause the oceans to freeze.

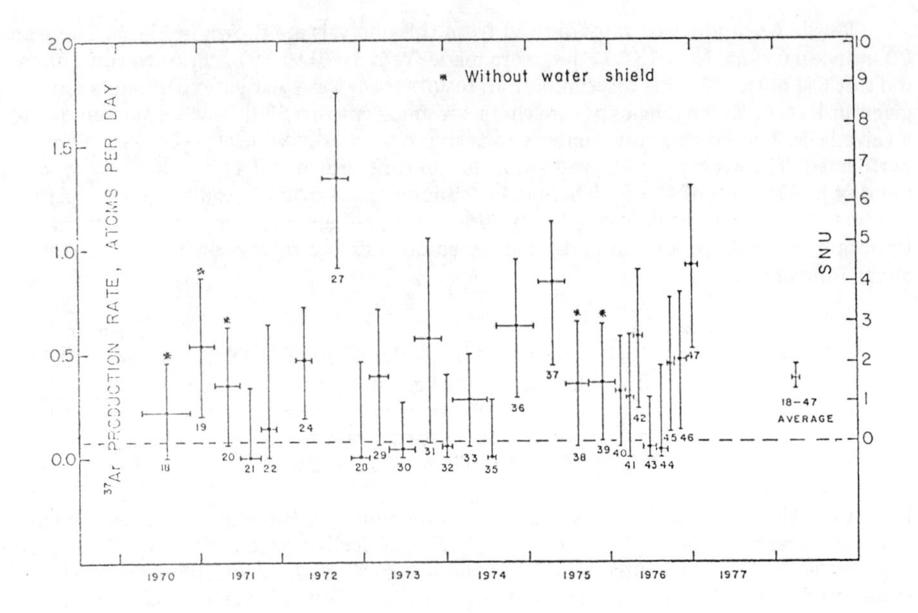

Fig. 1. Summary of results.

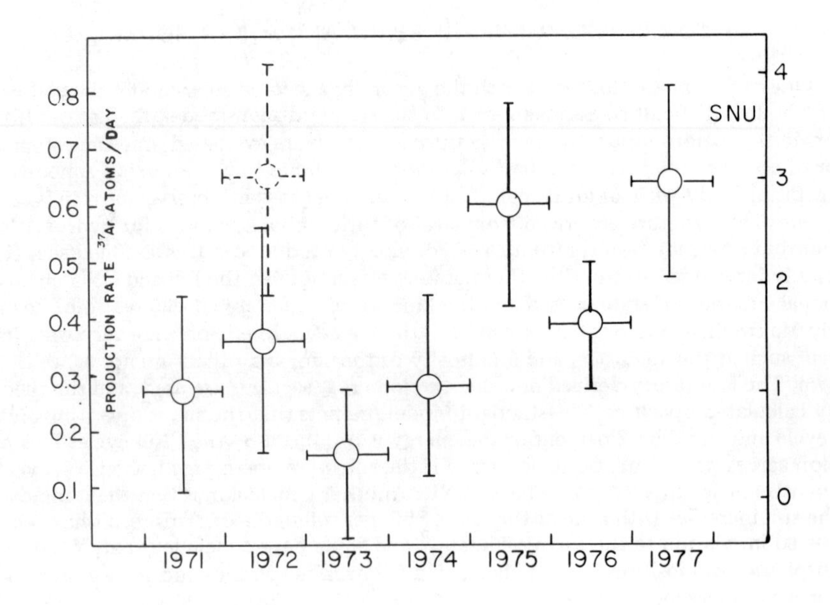

Fig. 2. Yearly averages.

During the last 10 years, almost all of the basic ideas of solar structure have been reexamined. Many effects, such as rapid internal rotation, periodic mixing, pure helium core, and intense magnetic fields, have been invoked in an effort to reduce the temperature in the central regions and thereby reduce the ^{8}B flux. These models are not satisfactory in that they are not consistent with observation, are not stable for long periods, or violate some concepts generally accepted in stellar evolution. There has been an extensive examination of the possibility that the sun is periodically mixed. However, as this question now stands, there is no satisfactory mechanism. Furthermore, to reduce the ^{8}B flux sufficiently to account for our observation requires mixing a large fraction of the interior of the sun. One of the most reasonable models that is consistent with our results is one in which the interior of the sun is essentially devoid of heavy elements. This reduces the opacity and allows radiation to escape more readily from the interior. The reduced central temperature results in a dramatic reduction in the ^{8}B flux. For this model to be convincing, one needs a mechanism for adding the heavy elements to the solar surface after the sun becomes a stable main-sequence star. Several mechanisms have been proposed, such as the infall of cometary-like debris or the collection of material by the sun during its travel around the galaxy. These various nonstandard models have been discussed in various review articles.[9] However, there is to date no critical review of all the aspects of the solar neutrino question.

There has been an extensive examination of the laboratory measurements of nuclear-reaction cross sections of specific interest to the P-P chain and possible variations. At the present time, the nuclear physicists feel content that all the questions of cross section values, possible resonances in critical reactions, and possible new nuclear reactions have been answered.[10] Neutrino properties are another question that has been with us for about 10 years. Since the travel time for the neutrino and the amount of matter that the neutrino must pass through are large for solar neutrinos, it is possible that neutrino decay, oscillations, and small scattering processes could affect the terrestrial flux.All of these possible processes have been considered.[9] Neutrino oscillations have been discussed, considering both vacuum oscillations[11] and matter oscillations.[12] If either of these oscillations occurs, it could severely alter the interpretation of any solar neutrino observation. In fact, this is an important consideration in our thinking about new solar neutrino experiments.

NEW EXPERIMENTS

A number of new experimental approaches have been proposed in recent years. The ^{37}Cl experiment is relatively easy and simple to carry out, and the target element is rather inexpensive. This will not be the case for the next solar neutrino experiment! In addition to the usual difficulties, there are some added requirements imposed by our theoretical interest. The rate of the initiating P-P reaction in the sun is essentially independent of the variations in the solar structure. All solar models forecast the same flux of these low-energy neutrinos (0-0.42 MeV). From the viewpoint of astrophysics, one has great confidence that these low-energy neutrinos are being produced in the sun at the calculated rate. This is an important consideration if one uses a solar neutrino experiment to test for neutrino oscillations. Needless to say, a solar neutrino experiment tests for oscillation lengths much greater than is possible with experiments at reactors or accelerators. With these considerations in mind, we favor an experiment that is capable of observing the P-P reaction neutrinos, though any experiment with sufficient sensitivity to observe any part of the solar neutrino spectrum would be very important. The ultimate goal is to determine the energy spectrum of neutrinos from the sun, and a way of obtaining this information is to use several radiochemical detectors with different thresholds. The ultimate technique would be a direct counting method that observes the energy of the neutrino and its direction.

If one examines all beta emitters with allowed or superallowed transitions and low disintegration energies that could be used for observing low-energy neutrinos, one finds only very few that are suitable. Table II is a list of the ones that are considered reasonably satisfactory and are now being considered in various laboratories. The table is divided into radiochemical detectors and direct counting neutrino detectors.

Let us first discuss the various radiochemical approaches. The reaction with the lowest threshold is the one with thallium in which ^{205}Tl (70.9%) captures a neutrino to form ^{205m}Pb, which rapidly decays to ^{205}Pb. The product ^{205}Pb has a very long half-life (1.6×10^7 yr), so it is necessary to use a very old mineral as the target material. Mel Freedman and his associates at Argonne National Laboratory propose using 3-10 kg of a mineral low in lead that has been exposed at depth underground.[13] There is some difficulty in obtaining the mineral, and there is at present an uncertainty in knowing the exact value of the cross section. Another similar case is the neutrino capture in ^{81}Br to form ^{81m}Br that decays to the long-lived ^{81}Kr (half-life 2.1×10^5 yr). The target material suggested is a salt deposit that has a small amount of bromine present.[14] These experiments have the unique ability of measuring the neutrino flux in the past, the thallium experiment measures the H-H reaction, and the bromine experiment could measure the ^{7}Be decay occurring in the sun. However, any experiment that uses a natural deposit can have serious built-in background effects.

A very attractive reaction is the one using gallium. It has a low threshold and, therefore, the dominant signal would come from the low-energy neutrinos from the H-H reaction. The product ^{71}Ge has a convenient half-life and its decay is relatively easy to observe in a gas-proportional counter using germane (GeH_4) as the counting gas. The chemical procedures for efficiently extracting ^{71}Ge from gallium metal or gallium chloride solution have been developed.[16] The major problem with this approach is to obtain the use of 50 tons of gallium for a few years. The material is produced on a sufficient scale, but it is expensive. Of course, it can be returned to the industrial market at the end of the experiment and thus recover the cost of the material. A solar neutrino detector based on the ^{7}Li(ν,e$^-$)^{7}Be reaction has many advantages. The neutrino-capture reaction has a relatively high cross section (superallowed). The threshold is slightly higher than chlorine, but because of the superallowed character of the transition this experiment would have a high sensitivity to the medium-energy neutrinos from the $H + H + e^- \rightarrow D + \nu$ reaction and from the decay of ^{13}N, ^{15}O, and ^{8}B. Because of the superallowed character of the ^{7}Li(ν,e$^-$)^{7}Be transition, the lithium experiment requires the least amount of material, only 5 tons for 1 capture per day (standard model)! The major difficulty with the lithium experiment is in measuring the ^{7}Be produced. There are several techniques that could be used, but as yet no really satisfactory method has been developed. Table III compares the relative sensitivity of the three radiochemical detectors using gallium, chlorine, and lithium as target material. Examining this table makes clear that the gallium detector responds mainly to P-P neutrinos, the chlorine detector responds primarily to the energetic neutrinos from ^{8}B, and the lithium detector has a more uniform response to all neutrino sources. If we had results from all three of these radiochemical detectors, there would be sufficient information to determine the solar neutrino spectrum. This is the goal of the Brookhaven program.

Table II. Proposed solar neutrino detectors.

Neutrino-capture reaction	Product half-life	Threshold energy (MeV)	Tons of element needed for 1 ν-capture/day (for the standard solar model), all sources
$\nu + {}^{205}Tl \rightarrow e^- + {}^{205m}Pb \rightarrow {}^{205}Pb$	1.6×10^7 years	0.048	13
$\nu + {}^{55}Mn \rightarrow {}^{55}Fe + e^-$	2.6 years	0.231	290
$\nu + {}^{71}Ga \rightarrow {}^{71}Ge + e^-$	11 days	0.233	38
$\nu + {}^{81}Br \rightarrow e^- + {}^{81m}Kr \rightarrow {}^{81}Kr$	2.1×10^5 years	0.490	660
$\nu + {}^{37}Cl \rightarrow {}^{37}Ar + e^-$ (present system)	35 days	0.814	603
$\nu + {}^{7}Li \rightarrow {}^{7}Be + e^-$	53 days	0.862	5
$\nu + {}^{115}In \rightarrow e^- + {}^{115m}Sn \rightarrow {}^{115}Sn + 2\gamma$	Direct counting	0.128	3.1
$\nu + D \rightarrow 2H + e^-$	Direct counting	~5	~6 (${}^{8}B$ flux only)
$\nu + e^- \rightarrow \nu + e^-$		~7	~2000 (${}^{8}B$ flux only)

Table III. Relative sensitivities of Ga, Cl, and Li detectors to the neutrino sources in the sun.[a]

Neutrino source	Neutrino energy (MeV)	Percentages of the total rate from each neutrino source in the sun (standard model)		
		$^{71}Ga(\nu,e^-)^{71}Ge$	$^{37}Cl(\nu,e^-)^{37}Ar$	$^{7}Li(\nu,e^-)^{7}Be$
$H + H \rightarrow D + e^+ + \nu$	0-0.42	71	0	0
$H + H + e^- \rightarrow D + \nu$	1.44 line	2	5	33
^{7}Be decay	0.861	23	17	11
^{13}N decay	0-1.20	1	1	4
^{15}O decay	0-1.74	2	3	15
^{8}B decay	0-14	<1	74	36
$\Sigma\phi\sigma$ in SNU		92	4.7	27.3

[a] All cross sections are from J. N. Bahcall (Ref. 16).

A direct observation of the neutrino interaction itself could give information on the energy of the neutrino and its direction. These features are only possible if the neutrino energy is relatively high. Direct counting experiments to observe solar neutrinos have been discussed for 15 years.[1,17] However, it is very difficult to reduce the background counting rate of a few-ton detector sufficiently low to observe the feeble signal from solar neutrinos. The only hope of success is to take advantage of a very unique signal from the neutrino interaction or to observe a neutrino interaction with energy release above that of background processes. Recently, R. S. Raghavan of Bell Labs has proposed using the neutrino capture in ^{115}In to produce an isomeric state in ^{115}Sn that rapidly decays (3.2 μsec) by emitting two successive characteristic gamma rays. This unique delayed triple-coincidence process could identify the neutrino-capture event sufficiently to distinguish the process from various background events. The reaction has a low threshold and could observe the neutrinos from the H-H reaction. A particular arrangement of indium-loaded liquid scintillation counters has been suggested,[18] but background effects must be carefully studied before feasibility can be clearly demonstrated. A detector based upon this reaction can in principle also measure the energy spectrum of low-energy neutrinos.

One of the early processes considered for observing ^{8}B neutrinos was the capture in deuterium producing an electron with an energy above 7 MeV. A detector was built by T. L. Jenkins (Case) about 10 years ago that used 2000 liters of D_2O, but various background processes limited its sensitivity. We know now that the ^{8}B flux is below 1×10^{6} cm^{-2} sec^{-1} from the chlorine experiment, so that observing ^{8}B neutrinos by this method is extremely difficult. Recently, A. Fainberg (Brookhaven-Syracuse) has proposed building a D_2O Čerenkov detector of high resolution.[19] His present aim is to study backgrounds, to determine if such a detector is capable of observing the low fluxes of ^{8}B neutrinos. A deuterium detector of this design is needed for observing pulses of neutrinos from collapsing stars. Present theories of stellar collapse predict an initial pulse of neutrinos of a few hundredths of a second duration followed by a continued pulse of neutrino-antineutrino pairs that may last many tens of seconds. A 10-30 ton D_2O Čerenkov detector of the type proposed by Fainberg is the best means of observing this sharp characteristic pulse from a supernova event. Such a detector could observe the constant flux of energetic solar neutrinos.

Neutrino-electron scattering also has been regarded as a promising means of observing energetic ^{8}B neutrinos.[1] Observing the scattering event by a sandwich detector system made of alternating layers of thick plastic-scintillator slabs and spark-chamber modules has been recently suggested by H. Chen of the University of California, Irvine.[20] Studies of background processes have been made with a pilot system at the LAMPF accelerator that indicate a detector of this design would have a sufficiently low background to allow observing the ^{8}B flux. A detector of this design with the ability of defining the direction of the scattered electron would identify the sun as the source of the neutrinos that are observed. These various direct counting experiments look promising and perhaps in a future neutrino '80-'90 conference the direct observation of solar neutrinos will be reported.

REFERENCES

1. F. Reines, Proc. R. Soc. London A301, 159 (1967).
2. J. N. Bahcall, W. F. Huebner, N. H. Magee, Jr., A. L. Merts, and R. K. Ulrich, Astrophys. J. 184, 1 (1973).
3. J. N. Bahcall, Astrophys, J. 216, L115 (1977).
4. R. Davis Jr., D. S. Harmer, and K. C. Hoffman, Phys. Rev. Lett 20, 1205 (1968); R. Davis, Jr., J. C. Evans, V. Radeka, and L. C. Rogers, in proceedings of the Neutrino '72 Europhysics Conference, Balatonfüred, Hungary, 1972 (OMKD Technoinform, Budapest, 1972), Vol. 1, p. 5; J. N. Bahcall and R. Davis, Jr., Science 191, 264 (1976).
5. R. Davis, Jr., Acta Phys. Acad. Sci. Hung. 29, Suppl. 4 (Eleventh International Conference on Cosmic Rays, Budapest, 1969, edited by A. Somogyi), 371 (1970); R. Davis, Jr., in Proceedings of the Conference on Astrophysical Aspects of Weak Interactions, Cortona, Italy, 1970 (1970), p. 59 [see also R. Davis, Jr., BNL Report No. BNL 14822 (unpublished)].
6. A. W. Wolfendale, E. C. M. Young, and R. Davis, Jr., Nature (Phys. Sci.) 238, 130 (1972); W. S. Pallister and A. W. Wolfendale, in Neutrinos '74, proceedings of the Fourth International Conference on Neutrino Physics and Astrophysics, Philadelphia, edited by C. Baltay (AIP, New York, 1974), p. 273; G. V. Domogatsky and R. A. Eramzhyan, Bull. Acad. Sci. USSR, Phys. Ser. 41, 169 (1977) [Isv. Akad. Nauk SSSR, Ser. Fiz. 41, 1969 (1977)]; E. L. Fireman, in Neutrino '77, proceedings of the International Conference on Neutrino Physics and Neutrino Astrophysics, Baksan Valley, U. S. S. R., 1977, edited by M. A. Markov (Nauka, Moscow, 1978), Vol. 1, p. 53.
7. R. K. Ulrich, in Neutrinos '74, p. 259.
8. R. T. Rood and M. Newman, Science 198, 1035 (1977).
9. J. N. Bahcall and R. Sears, Annu. Rev. Astron. and Astrophys. 10, 25 (1972); J. N. Bahcall, in Proceedings of the International Conference on Nuclear Physics, Munich, 1973, edited by J. de Boer and H. J. Mang (North Holland, Amsterdam, 1973), Vol. 2, p. 681; B. Kuchowicz, Rep. Prog. Phys. 39, 291 (1976); R. Davis, Jr., and J. C. Evans, Jr., in New Solar Physics, edited by J. A. Eddy (AAAS Selected Symposium Series No. 17, Westview, Boulder, Colorado, 1978).
10. P. D. Parker, Yale University, Wright Nuclear Structure Laboratory Report No. 3074-287 (unpublished); W. R. Fowler, LAMPF Newsletter 10, No. 1, p. 32 (1978).
11. V. Gribov and B. Pontecorvo, Phys. Lett 28B, 493 (1969); J. N. Bahcall and S. C. Frautschi, Phys. Letters 29B, 623 (1969); A. K. Mann and H. Primakoff, Phys. Rev. D 15, 655 (1977); H. Primakoff, in Neutrinos-78, proceedings of the International Conference on Neutrino Physics and Neutrino Astrophysics, West Lafayette, Indiana, 1978, edited by E. C. Fowler (Purdue University, 1978), p. 995; A. K. Mann, Chap. 3 of the present proceedings.
12. L. Wolfenstein, Phys. Rev. D 17, 2369 (1978); Neutrinos-78, p. C3; and article in these proceedings.
13. M. S. Freedman et al., Science 193, 1117 (1976); M. S. Freedman, in Proceedings of the Informal Conference on the Status and Future of Solar Neutrino Research, Brookhaven National Laboratory, 1978, edited by G. Friedlander (BNL Report No. BNL 50879, 1978), Vol. 1, p. 313.
14. R. D. Scott, Nature 264, 729 (1976); T. Kirsten and W. Hampel (private communication) and T. Kirsten, in Proceedings of the Informal Conference on Solar Neutrinos (BNL 50879), Vol. 1, p. 305.

15. J. N. Bahcall, B. T. Cleveland, R. Davis Jr., I. Dostrovsky, J. C. Evans, Jr., W. Frati, G. Friedlander, K. Lande, J. K. Rowley, R. W. Stoenner, and J. Weneser, Phys. Rev. Lett. 40, 1351 (1978).
16. J. N. Bahcall, Rev. Mod. Phys. 50, 881 (1978).
17. F. Reines, Proceedings of the International Conference on Neutrino Physics and Neutrino Astrophysics, Moscow, 1968 (Moscow, 1969), Vol. 2, p. 129; R. Davis, Jr., Proceedings of the Solar Neutrino Conference, Irvine, California, 1972, edited by F. Reines and V. Trimble (University of California, Irvine, 1972), p. A-14.
18. R. S. Raghavan, Phys. Rev. Lett. 37, 259 (1976); L. Pfeiffer, A. P. Mills, Jr., R. S. Raghavan, and E. A. Chandross, Phys. Rev. Lett. 41, 63 (1978).
19. A. M. Fainberg, in Proceedings of the Informal Conference on Solar Neutrinos (BNL 50879), Vol. 2, p. 93.
20. H. H. Chen, University of California, Irvine, Report UCI-Neutrino-No. 14, 1975 (unpublished); Proceedings of the Informal Conference on Solar Neutrinos (BNL Report No. 50879), Vol. 2, p. 55.

SOLAR NEUTRINOS AND THE CATALYTIC ROLE OF A THIRD PARTICLE IN HYDROGEN BURNING

Yu. S. Kopysov
Institute for Nuclear Research, Academy of Sciences of the USSR, Moscow, USSR
(Reported by G. T. Zatsepin)

ABSTRACT

The possibility of catalytic acceleration of the proton-proton reaction in the solar interior is considered. It is shown that such acceleration may well be the case when two protons beta-decay to form a deuteron in the vicinity of a third nuclear particle, such as p, ^4He, ^3He, and d. The effect of nuclear catalysis on the solar neutrino flux is also discussed.

As is known, an increase in the rate of the reaction

$$p + p \rightarrow d + e^+ + \nu \qquad (1)$$

in the solar interior gives rise to a sharp decrease in the flux of high-energy solar neutrinos from the ^8B decay.[1-4] In view of the persisting disparity between theoretical expectations of the solar neutrino flux and the observational results of the Brookhaven solar neutrino experiment,[5] the search of catalytic acceleration of the fusion of two protons to form a deuteron is of particular interest. In particular, one may suppose that a third particle, such as p, d, ^3He, or ^4He, would accelerate the fusion of protons giving rise to a catalyzed version of reaction (1), such as

$$p + p + p \rightarrow d + p + e^+ + \nu\,, \qquad (2)$$

$$p + p + d \rightarrow d + d + e^+ + \nu\,, \qquad (3)$$

$$p + p + {}^3He \rightarrow d + {}^3He + e^+ + \nu\,, \qquad (4)$$

or

$$p + p + {}^4He \rightarrow d + {}^4He + e^+ + \nu\,, \qquad (5)$$

respectively.

As a general rule, the rate of a three-body reaction is less than the rate of a two-body one. However, if the fusion of three particles in reactions (2) - (5) should proceed via the formation of a compound nuclear complex, such as ^3Li, $^4He^*$, ^5Be, or ^6Be, respectively, the rate of each of the reactions (2) - (5) may prove to be larger than the rate of two-body reaction (1).

The important condition for nuclear catalysis is the proximity of the energy E_c of a compound nucleus to the thermal energy kT of the initial particles, as well as the smallness of the beta-decay width Γ_f, which must obey the inequality

$$\Gamma_f \ll \Gamma_i, \qquad (6)$$

where Γ_i is the entrance width of a nuclear complex.

Under condition (6), the ratio of the rates of the proton-proton reaction induced by the particle "a", W_{ppa}, and of the two-body reaction (1), W_{pp}, can be written as

ISSN:0094-243X/79/520028-04$1.50

$$\frac{W_{ppa}}{W_{pp}} = \sqrt{\frac{\pi}{8}}\,\frac{g_c A_c^{3/2}}{g_a A_a^{5/2}}\,\frac{\rho X_a}{M}\,\frac{\Gamma_f}{S(E_o)}\,\frac{kT}{\Delta E_o}\left(\frac{2\pi\hbar^2 c^2}{Mc^2 kT}\right)^{5/2} \exp\left(\frac{3E_o - E_c}{kT}\right), \tag{7}$$

where g_a, g_c and A_a, A_c are the statistical and atomic (in units of the proton mass M) weights of the nuclei "a" and "c", respectively;

$$\Delta E_o = 4\left(\frac{1}{3} kTE_o\right)^{1/2},\ E_o = \left\{\frac{1}{4}(kT)^2 E_G\right\}^{1/3},\ E_G = \left(\frac{\pi e^2}{\hbar c}\right)^2 Mc^2,$$

$$S(E_o) = 3.78 \times 10^{-46}\left\{1 + 0.417\left(\frac{kT}{3E_o}\right) + 12.6\left(\frac{kT}{3E_o}\right)^2 + 36.6\left(\frac{kT}{3E_o}\right)^3\right\} \text{keV cm}^2; \tag{8}$$[6]

ρ is the mass density; and X_a is the mass fraction of the catalyzing nuclei "a" in solar matter.

Let us consider reaction (2). If we assume the quantum numbers of the nuclear complex ^3Li to be $J^\pi = \frac{3^-}{2}$, then $g_c = 4$. Putting in (7) $A_c = 3$ and

$$\Gamma_f = \gamma \times 10^{-22} \text{ keV}, \tag{9}$$

where 10^{-22} keV is the beta-decay width of hypothetical ^2He on the assumption that the radial nuclear matrix element for beta decay of ^2He equals unity, we find at $T = 10^7$K and $\rho X = 60$ g cm^{-3}, which are representative for the standard solar model, that

$$\frac{W_{ppp}}{W_{pp}} = 1.18 \times 10^4\ \gamma \times 10^{-0.504\ E_c\ (\text{keV})}. \tag{10}$$

Let γ in (9) be equal to unity. Then, according to equation (10), the rate of reaction (2) will become higher than the rate of reaction (1) when $E_c \lesssim 8$ KeV, and the enhancement factor will be 10^3 - 10^4 when $E_c \lesssim 1$ keV. Thus, in principle, strong catalytic action may be available. In order to solve the solar neutrino problem, it is sufficient for the enhancement factor to be equal to 2 or 3. Therefore, it is plausible that the radial nuclear matrix element and width Γ_f of the beta-decaying compound nucleus could be smaller to the extent needed.

The nucleus ^3Li has never been observed. Is there any reason to suspect its existence?

The emergence of a ^3Li quasistationary state (presumably with $J^\pi = 3/2^-$) is much facilitated owing to the existence of two protons in a virtual s-state with energy ~700keV. It is not necessary for a third proton to have an independent level in the field of ^2He and it is sufficient for the virtual level of ^2He to be lowered by the nuclear field of a third thermal proton down to the energy range below 8 keV. The virtual state of ^2He, which was initially a two-particle one, becomes by inclusion of a self-consistent interaction of three particles a quasistationary low-lying state of three protons, which may explain in a simple way the formation of the nuclear complex ^3Li. ^2He, which is unbound by ~700 keV against decay to p+p, can be compared to activated complexes in chemical reactions with activation energy ~700keV. In these terms, a catalytic action of a third particle can be recognized as a decrease in the activation energy of the pp reaction.

The size of such an unusual nucleus as ^{3}Li is about 10^{-11} cm. It is difficult to produce it by reactions among ordinary compact nuclei because the probability for all three protons of such a friable system to be simultaneously in the range of nuclear forces is no more than 10^{-5}, but may be far less because of Coulomb and centrifugal barriers. The most promising way to search for reaction (2) appears to be by experiments with a dense, high temperature, pure hydrogen laser plasma. This reaction might be identified by means of two γ-quanta characteristic of the positron annihilation as well as by an anomalous (as compared with reaction (1)) dependence of a γ-quantum yield on density and temperature.[7] One may also endeavor to search for ^{3}Li by means of reactions such as $^{3}He + {}^{3}He \rightarrow {}^{3}H + {}^{3}Li$, $^{10}B + {}^{3}He \rightarrow {}^{10}Be + {}^{3}Li$, $^{7}Be + {}^{3}He \rightarrow {}^{7}Li + {}^{3}Li$, etc., near the threshold of its formation, but it is essential to bear in mind that such efforts may fail because of the unusual structure of ^{3}Li.

The neutrino spectrum from reaction (2) has to be somewhat softer than that of reaction (1). Hence the solar neutrino counting rate in a detector which uses ^{71}Ga as a target may turn out to be less than the expected one.

All that has been said about reaction (2) applies equally well to reactions (3) - (5). Reaction (2) and the chain (4) + ($d + p \rightarrow {}^{3}He + \gamma$) are of special interest in solar physics. These reactions are autocatalytic, and thanks to this property they could cause a tendency for the sun to form a dissipative structure. However, in view of the low concentrations of d and ^{3}He in regions of hydrogen burning, their role may be significant only in those regions where the burning rates of d and ^{3}He are small in comparison with the rates of reactions (3) and (4), respectively. The catalytic action of d, ^{3}He, and ^{4}He can be investigated in a laser plasma as well. Moreover, one may search for the nuclear complexes in question by all nuclear reactions which come to mind. A priori, it is unknown which path will lead one to positive results. Keeping in mind the far reaching analogy with chemistry, let us remember that chemical catalizers are usually selected by empirical methods.

According to our conception of friable nuclear complexes, the radii of the compound nuclei like ^{3}Li, $^{4}He*$, ^{5}Be and ^{6}Be must be so large that the corresponding Coulomb barriers in reactions such as (2) - (5) may prove to be anomalously low and such reactions may be dominant at low temperatures. By means of such reactions, one may hope to construct a low-temperature solar model like that put forward by A. B. Severny et al.[8] to account for solar oscillations with a period of 2 h 40 min.

In conclusion, one must note that some indication of the existence of the nuclear complex ^{6}Be lying below the ground state of ^{6}Be may be found in Ref. 9, in which the reaction ^{6}Li (^{3}He, ^{3}H) ^{6}Be was investigated.

The author thanks Prof. G. T. Zatsepin for discussions and stimulating interest in this work. Thanks are also due to I. R. Barabanov, E. V. Bugaev, P. N. Vasil'ev, V. N. Gavrin, G. V. Domogatsky, I. M. Zheleznykh, A. A. Komar, Yu. Ya. Markov, V. N. Fetisov, and Prof. A. E. Chudakov for useful discussions.

REFERENCES

1. A. Finzi, Astrophys. J. 189, 157 (1974).
2. M. J. Newman and W. A. Fowler, Phys. Rev. Lett. 36, 895 (1976).
3. Yu. M. Andreev, E. V. Bugaev, and Yu. S. Kopysov, Zh. Eksp. Teor. Fiz. Pis'ma 25, 593 (1977).
4. I. R. Barabanov, A. I. Egorov, V. N. Gavrin, Yu. S. Kopysov, and G. T. Zatsepin, in Neutrino '77, proceedings of the International Conference on Neutrino Physics and Neutrino Astrophysics, Baksan Valley, U.S.S.R., 1977, edited by M. A. Markov (Nauka, Moscow, 1978), Vol. 1, p. 20.
5. J. K. Rowley, B. T. Cleveland, R. Davis, Jr., and J. C. Evans, in Neutrino '77, Vol. 1, p. 15.
6. J. N. Bahcall and R. M. May, Astrophys. J. Lett. 152, L17 (1968).
7. Yu. S. Kopysov and V. N. Fetisov, Krat. Soob. po Fiz. (Short Commun. on Phys.) 10, 60 (1972).
8. A. B. Severny, V. A. Kotov and T. T. Tsap, Nature 259, 87 (1976).
9. D. F. Geesaman, R. L. McGrath, P. M. S. Lesser, P. P. Urone, and B. VerWest. Phys. Rev. C 15, 1835 (1977).

THE HOMESTAKE LONG-RANGE NEUTRINO DETECTOR RESEARCH PROGRAM*

M. Deakyne, W. Frati, K. Lande, C. K. Lee, and R. I. Steinberg
Physics Department, University of Pennsylvania, Philadelphia, Pennsylvania 19174

E. Fenyves
University of Texas at Dallas, Richardson, Texas 75080

ABSTRACT

The research program of the Homestake long-range neutrino detector is described. The main elements of this program are to look for ν_e and $\bar{\nu}_e$ bursts from the gravitational collapse of massive stars, search for high-energy neutrinos from localized astronomical sources, investigate possible vacuum or matter oscillations of cosmic-ray neutrinos as they traverse the earth, and serve as a potential target for directed beams of neutrinos from high-energy particle accelerators.

INTRODUCTION

We are presently embarked on the construction of a large counter-hodoscope neutrino detector[1] that is located at a depth of 4400 meters water equivalent in the Homestake Gold Mine in Lead, South Dakota. This hodoscope is in the form of a closed, hollow box, 20 meters long, 10 meters wide, and 7 meters high, that surrounds the Brookhaven ^{37}Cl Solar Neutrino Detector. The counter elements that form the sides and bottom of the hodoscope are water Čerenkov counters 2 m × 2 m × 1.2 m and those that form the top of the hodoscope are liquid scintillation counters that are 1.5 m × 1.5 m. Forty percent of the hodoscope elements are now in place and operating and we expect the entire detector to be functioning by the end of 1978 (Fig. 1).

This apparatus is designed to carry out a number of investigations in astronomy, astrophysics and elementary particle physics. These include:

1) Look for bursts of neutrinos and antineutrinos emitted in the final gravitational collapse of massive stars[2];
2) Search for localized astronomical sources of high-energy neutrinos;
3) Determine whether neutrinos produced by the interactions of cosmic-ray primaries in the atmosphere oscillate as they pass through the earth for path lengths between 10^2 km and 10^4 km;
4) Investigate the production characteristics of energetic muons ($E > 10^{13}$eV) made by very-high-energy cosmic-ray primaries;
5) Set improved limits on the proton lifetime;
6) Measure the flux of charged cosmic-ray secondaries passing through the Brookhaven Solar Neutrino Detector;
7) Provide a potential detector for directed neutrino beams from high-energy particle accelerators.

*This research was supported in part by the U.S. Department of Energy and the National Science Foundation.

ISSN:0094-243X/79/520032-06$1.50

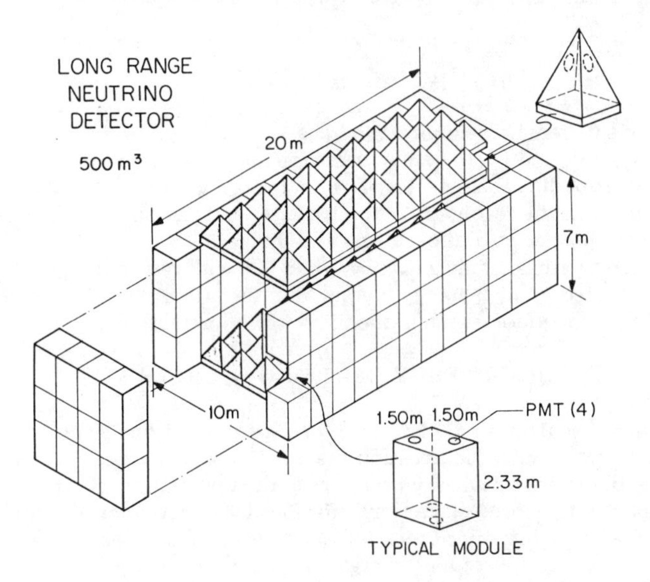

Fig. 1. Drawing of the Long Range Neutrino Detector showing the location of the H_2O Čerenkov counter modules that form the sidewalls and floor of the detector and the scintillation counters that form the roof.

NEUTRINO DETECTION

Low energy ν_e and $\bar{\nu}_e$ are detected by this apparatus by the reactions

$$\nu_e + e^- \rightarrow \nu_e + e^-,$$
$$\bar{\nu}_e + p \rightarrow n + e^+ ,$$

in the water fill of the Čerenkov-counter modules.

For high energy ν_μ and $\bar{\nu}_\mu$, the main target for the production of μ^- and μ^+ secondaries is the rock surrounding the detector. The neutrino flight direction is determined by the hodoscope elements the secondary muon traverses and the time of flight between these elements. The angular resolution of this system is $\lesssim 5°$.

For muon flight paths making angles less than 70° with respect to the vertical, the signal will be dominated by high-energy muons from the atmosphere. For angles between 80° and 180°, the muon signal is due to neutrino interactions.

HIGH ENERGY NEUTRINO SOURCES

The high-energy neutrinos have two basic origins: one is from extraterrestrial, possibly localized, astronomical sources and the other is from the decay of high-energy π and K mesons produced by cosmic-ray primaries in the upper atmosphere.

There have been numerous suggestions[3] for localized astronomical sources of high-energy neutrinos. Although many of these possibilities are quite speculative, it seems likely that the production sites of high-energy charged cosmic-ray primaries also emit neutrinos. Except for energies $>10^{18}$ eV, these charged-particle cosmic rays have their flight directions completely randomized by the interstellar magnetic fields of the galaxy and so provide no information about the location of their source. Since the neutrinos are uneffected by magnetic fields, they will travel in straight lines from production site to detector and, thus, permit source identification. We have already initiated such a neutrino sky survey and, as further elements of the telescope are brought into operation, we will continue this survey with increasing sensitivity.

The production of neutrinos in the upper atmosphere in the 1-100 GeV range is well understood, since these neutrinos are produced together with muons in the decay of π and K mesons. In the energy range above 100 GeV, the π and K-decay contributions diminish and it is possible that other, "direct", production mechanisms become significant. In the absence of new neutrino phenomena and after correction for angular dependence of π and K decay probabilities, these cosmic ray neutrinos provide an isotropic source of muon secondaries for our detector for $80° \leqslant \theta \leqslant 180°$. These neutrinos traverse the Earth with flight paths between 10^2 km and 10^4 km and pass through matter with $\int \rho dx \lesssim 10^{10}$ gm/cm^2.

NEUTRINO OSCILLATIONS

One question basic to the detection of neutrinos from distant sources is the behavior of these particles during the source-detector transit. Present accelerator-neutrino studies do not indicate any significant effects on neutrinos in the passage through massive bodies (i.e., the earth) or on long flight paths. There are, however, suggested phenomena which might not be observable in the experiments done to date. A particularly interesting suggestion is that neutrinos oscillate in a fashion analogous to K^0 regeneration. One of these processes, vacuum oscillation,[4] depends on the mass differences between the various neutrino states, while the other, matter oscillation,[5] would result from a small lepton nonconserving neutral current interaction in the elastic scattering of neutrinos from matter. The existence of both of these processes would result in an interference phenomenon.

The long-path length transmission of neutrinos through the Earth provides an ideal arrangement for searching for these oscillations. They would manifest themselves by a variation of neutrino-induced muon flux as a function of angle of flight path with the vertical (Fig. 2).

For pure vacuum oscillations, we have an oscillation length[6]

$$L = 2.5 \frac{E_\nu \text{ (MeV)}}{(m_{\nu_1}^2 - m_{\nu_2}^2)\text{ (eV)}^2} \text{ meters.}$$

Using the earth's diameter, 1.2×10^7 meters, and a typical neutrino energy of 2×10^4 MeV we find that for

$$\Delta m^2 \geqslant 4 \times 10^{-3} \text{ eV}^2$$

our detector will be able to observe one or more full oscillations. In the range

$$4 \times 10^{-3} \text{ eV}^2 \geqslant \Delta m^2 \geqslant 10^{-3} \text{ eV}^2,$$

a sufficient fraction of an oscillation will occur to make the effect apparent.

The description of matter oscillations is more complex in that it depends on the amplitude and form of the lepton nonconserving term in the neutral-current interaction and whether or not vacuum oscillations simultaneously exist.[7]

The basic oscillation length is given by

$$L = \frac{2\pi}{GN_e} = \frac{2.7 \times 10^7}{N_e/6 \times 10^{23} \text{ (cm}^{-3})} \text{ meters ,}$$

where G is the Fermi constant and N_e is the number of electrons per unit volume. Since neutrino trajectories through the earth's diameter have $\langle N_e \rangle / 6 \times 10^{23} \approx 4$, several matter oscillations could occur.

The effect of the earth's structure, a dense core, $\langle \rho \rangle = 12$ gm/cm^3 for $R < 3500$ km, and a mantle with $\langle \rho \rangle = 4.5$ gm/cm^3, gives rise to rapid variations in $\int N_e \, dx$ for neutrino trajectories that pass through the earth's core. Indeed, matter oscillations of neutrinos, if they exist, might eventually be useful in exploring the density distribution of the central regions of the earth.

NEUTRINOS FROM HIGH ENERGY ACCELERATORS

There have been several suggestions for transmitting a beam of neutrinos from a high-energy particle accelerator to a distant detector. Our detector, although modest in size for such a program, could provide a detector for an initial attempt at such transmissions. It also can serve as a model of a much larger underground detector which can provide an increased sensitivity for long-distance neutrino transmission.

It would be interesting to compare a distant neutrino detector, such as ours, with the conventional nearby accelerator-sited detectors. An accelerator-neutrino beam consists of the following elements:

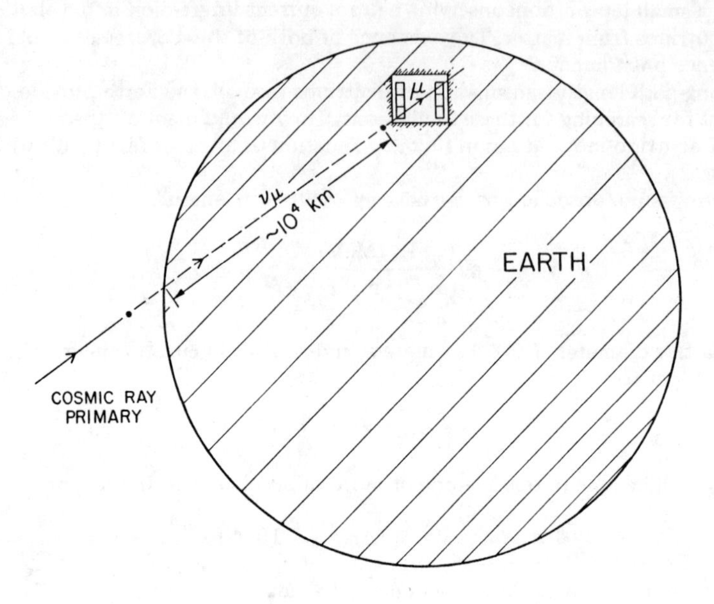

Fig. 2. Schematic view of a typical neutrino trajectory for the neutrino-oscillation experiment.

1. A π and K production target.
2. A "horn" or other focussing system which collects the π and K into a parallel beam.
3. A decay region in which the π and K decay into neutrinos.
4. A neutrino flight path and muon absorber followed by the neutrino target and detection system.

The primary distinctions between nearby and distant detection experiments are

1. For nearby detectors, the decay region is very short compared to the π and K mean free path for decay. Since $\Lambda_\pi = 7.5\ \gamma_\pi$ meters and $\Lambda_K = 3.6\ \gamma_K$ meters, for $\langle p_{\pi,K} \rangle = 250$ GeV/c, we have $\Lambda_\pi = 13$ km and $\Lambda_K = 1.8$ km. At Fermilab and at the CERN SPS the decay region is ≈ 300 m, so that a distant detector could have 30-40 times the initial neutrino flux available to nearby detectors.
2. The neutrino beam angular divergence due to the π and K decay kinematics is $(1/2)\ \gamma^{-1}_{\pi,K}$. For $\langle p \rangle = 250$ GeV/c this is about ¼ mrad for π decays and 1 mrad for K decays. At 1000 km from the accelerator this gives a spot size of 250-m radius. Since our detector has surface of 20 m $\times$ 7 m, we would intercept $\sim 10^{-3}$ of the incident neutrino beam.
3. For a distant underground detector, the surrounding rock serves as a neutrino target providing a thicker, and so more productive target than is usually used in nearby detectors.

In comparing the present Homestake detector with conventional accelerator-sited detectors, we find the relative neutrino-detection rates to be 30 (decay path) $\times\ 10^{-3}$ (detector aperture) $\times$ target-thickness factor. A "stretched" version of the present Homestake detector, with a surface area ≈ 25 times larger, would have a neutrino-detection rate comparable to that of present accelerator-sited detectors.

REFERENCES

1. M. Deakyne, W. Frati, K. Lande, C. K. Lee, R. Steinberg, E. Fenyves, and O. Saavedra, Neutrino '77, proceedings of the International Conference on Neutrino Physics and Neutrino Astrophysics, Baksan Valley, U.S.S.R., 1977 (Nauka, Moscow, 1978), Vol. 1, p. 170.
2. D. Z. Freedman, D. N. Schramm, and D. L. Tubbs, Annu. Rev. Nucl. Sci. 27, 167 (1977).
3. D. Eichler, Astrophys. J. 222, 1109 (1978).
4. B. Pontecorvo, Sov. Phys — JETP 53, 1771 (1967).
5. L. Wolfenstein, Phys. Rev. D 17, 2369 (1978).
6. A. K. Mann and H. Primakoff, Phys. Rev. D 15, 655 (1977).
7. L. Wolfenstein, Chap. 3 of these proceedings.

HIGH-ENERGY NEUTRINOS OF EXTRATERRESTRIAL ORIGIN

David Eichler
Department of Physics and Astronomy, University of Maryland,
College Park, Maryland 20742

ABSTRACT

Many astrophysical systems that exhibit nonthermal behavior — supernova remnants, pulsars, active galactic nuclei and quasars — may be detectable by DUMAND as point sources of high energy neutrinos.

INTRODUCTION

The two main questions of neutrino astronomy are 1) How are neutrinos produced in astrophysical systems, and 2) How can they be detected? Current schemes for detecting high-energy neutrinos will be discussed at length by other speakers at this conference.[1] For the purpose of my discussion, I will assume that a flux from a point source of order 10^2 eV cm^{-2} sec^{-1} in neutrinos more energetic than 3×10^{11} eV constitutes a detectable signal for a 10^9 ton detector in one year of viewing time. (This flux produces ten counts per year per 10^9 tons.) Signals as small as 10 eV cm^{-2} sec^{-1} may prove to be detectable. This flux level is not large by astrophysical standards: the brightest nonthermal sources (e.g. pulsars and active galactic nuclei) have fluxes far in excess of this value, so if they can somehow put even a small fraction of their total luminosity in the form of high-energy neutrinos, they are in principle detectable.

The production of high-energy neutrinos in astrophysical systems entails first the production of high-energy protons, followed by their degradation via pion-producing collisions against other protons. The most important reaction is

$$p + p \rightarrow p + p + \pi^{+}s + \pi^{-}s + \pi^{o}s. \quad (1)$$

While the neutral mesons decay into gamma rays, the charged pions decay into three neutrinos and an electron (or positron) via the standard decay sequence. The total cross section for proton-proton collisions is roughly 40 mb, only weakly dependent on the collision energy, and about half the energy goes into pions in any given collision. Thus protons are stopped over a path length of about 30 gm cm^{-2}. The leading pion may contain a significant fraction of the collision energy; thus, a large fraction of the collision energy may be continued in a single neutrino.

Pions can also be produced in proton-photon collisions when the photon has an energy more than 600 MeV in the frame of the proton. The total cross section for this reaction depends on the collision energy and at maximum is

$$\sigma_{p\gamma} \sim 2 \times 10^{-28} \text{ cm}^2. \quad (2)$$

To produce a significant quantity of high-energy neutrinos, a system must a) produce fast particles efficiently, and b) contain enough target matter near the source of the fast protons to stop a significant fraction of them. Table I lists a number of astrophysical situations in which both these conditions may be met. The remainder of the talk will consist of more detailed discussion of such systems.

ISSN:0094-243X/79/520038-05$1.50

Table I. Possible high-energy neutrino sources.

Site of HE particle production	Target material	Maximum expected counting rate from typical source
Expanding supernova remnants	Ejecta, swept up matter, dense cloud from which star formed	10^2/yr-10^9 tons
Stellar explosions	Binary companion	$10^8/10^9$ tons per event
Pulsars	Binary companion, accreting matter young supernova remnant	10^3/yr-10^9 tons
Seyferts, radio galaxies, and QSOs	Clouds in nucleus, matter accreting onto compact object	10^3/yr-10^9 tons

SUPERNOVAE

Standard cosmic-ray models assume that a typical supernova puts out 6×10^{50} ergs in relativistic cosmic rays, with the spectrum proportional to $E^{-2.5}$. The neutrino counting rate in a 10^9 ton detector is then[2]

$$R \sim 10^{-3} n D^{-2} \text{ per day,} \tag{3}$$

where D is the distance in Kpc to the supernova remnant and n is the ambient density in atoms per cm^3 of the medium in which the remnant is situated. Regardless of the detailed mechanism by which the cosmic rays are accelerated in either the remnant or the shock itself, the basic fact remains that if the supernova occurs in a dense cloud where n may be as large as 10^3, it would then be a detectable point source of high energy neutrinos. Since supernovae are presumably rapidly evolving massive stars, it is reasonable to assume that many go off while still in the immediate vicinity of the cloud out of which the presupernova star formed. In addition, it is also likely that a presupernova star will undergo significant mass loss and thus create its own circumstellar cloud.

A supernova in a dense environment would also be detectable by gamma-ray satellites in the energy range $E_\gamma < 1$ GeV. Indeed, coincidences have been reported[3] between unidentified gamma-ray sources detected by the Cos B satellite and radio-emitting supernova remnants. The signal detected by Cos B, if assumed to be the low-energy end of an $E_\gamma^{-2.5}$ gamma-ray spectrum originating from π_0 decay, would yield 10 to 10^2 eV cm^{-2} s^{-1} in 10^{12}-eV gamma rays, and a comparable flux in high-energy neutrinos.

PULSARS

Pulsars accelerate high-energy protons and, if these protons collide with nucleons, neutrinos will be produced. Berezinsky[4] has pointed out that young pulsars, blanketed by a supernova remnant less than a year old, could be bright sources of high-energy neutrinos. However, their behavior at such early times is highly uncertain. Regardless of these uncertainties, it is reasonable to believe that a pulsar in a binary system could be a source of high-energy neutrinos for the first few thousand years of its existence.[5] The stellar wind of the companion could have a column density as high as 1 to 10 gm cm^{-2}. Its corona could provide further target material and an accretion cloud could conceivably surround the pulsar. If the companion is a red giant, its envelope could block nearly half the high-energy protons emitted by the pulsar. A pulsar emitting (roughly the luminosity of the Crab pulsar) 10^{37} ergs s^{-1} in high-energy particles ($E_p > 10^{13}$ eV) at a distance of D Kpc produces

$$R = 10^2 \eta D^{-2} \text{ counts/day,} \tag{4}$$

where η is the fraction of the luminosity converted to high-energy neutrinos by the target material near the pulsar. A typical value of η might be from 10^{-3} to ~ 1.

Helfand and Tademaru[6] have speculated, on observational grounds, that a significant fraction of all pulsars form in binaries. Stepanian[7] and co-workers have detected a flux in high energy gamma rays ($E_\gamma > 3 \times 10^{12}$eV) of about 60 eV cm^{-2} sec^{-1} from Cyg x 3, which is generally held to be a (heavily blanketed) pulsar in a binary system. A comparable flux in high energy neutrinos would be detectable with a 10^9 ton detector in perhaps a year or two of viewing time. Any process, such as Colgate's[8] shock mechanism, which produces high-energy particles promptly during the initial supernova explosion, will produce a prompt burst of neutrinos if the exploding star has a binary companion. For example, in the Type I supernova model where a white dwarf accretes beyond the Chandrasekhar limit from a red giant companion, the explosion will of course occur in the vicinity of that companion. The narrow pulse detected at earth would contain a maximum of

$$N = 10^8 \eta' D^{-2} (U/10^{48} \text{ ergs}) \text{ counts}, \tag{5}$$

where U is the energy output of relativistic particles, and η' is the fraction of them stopped by the binary companion and surrounding material. A typical value here for η' might be as high as 1/2.

GALACTIC NUCLEI

The most potent sources of high-energy neutrinos may be active galactic nuclei,[9] such as those of radio galaxies, Seyferts, and possibly Quasars. Such nuclei are capable of luminosities as high as 10^{43} to 10^{47} ergs $\sec^{-1}$. Since the radiation appears to be of nonthermal origin, it is believed that the power outputs in high-energy particles take on similar values. Since these high energy particles would be produced within the innermost regions of the nucleus, which have column densities well in excess of a typical galactic disk, the high-energy protons produced there might be expected to be stopped via nuclear collisions before escaping from the nucleus. (Consider, for example, that cosmic rays in our own galactic disk go through a few grams per cm^2 of matter before escaping. Were the column density of the disk a factor of 30 higher, the dominant loss mechanism would be nuclear collisions.) Moreover, it can be argued[9] that the accretion column surrounding a collapsed object in the nucleus may have column densities of 10^2 to 10^3 gm cm^{-2}, which would be sufficient to even stop the gamma-rays, but the neutrinos would still escape easily.

Thus, one could reasonably expect that such active nuclei put out a significant fraction ($\gtrsim 10\%$) of their entire luminosity in the form of high energy neutrinos. The nearby peculiar galaxy M82 (often thought to be an exploding galaxy because of its optical appearance) could give a neutrino count rate in excess of 1 per day if, indeed, there are nonthermal processes occurring in its nucleus. Other nearby active nuclei, such as Cen A and M87 could give count rates on the order ten per year. Grindlay and co-workers have reported[10] a high-energy ($E_\gamma > 3 \times 10^{11}$ eV) gamma-ray flux of order 50 eV cm^{-2} $\sec^{-1}$ from Cen A. A comparable flux in high-energy neutrinos would be detectable within a year of viewing time. (Remember that the atmospheric background per sq. deg. in a 10^9 ton detector is ~ 0.1 counts/yr; thus, as few as 0.3 counts/yr from the same sq. deg. would indicate a point source.)

Compact radio sources emit in bursts which almost certainly originate within regions smaller than a light day. Thus, a few years worth of neutrinos, as computed from average luminosities, could arrive within a few hours, making them much more easy to detect. It has been suggested[11] that as many as 10^4 counts per 10^9 tons per burst could be recorded from a radio burst, for example, the nucleus of 3C 120.

Particularly significant is the fact that $\sim 10\%$ of the flux from Cen A is in the form of high-energy gamma rays. This corroborates the picture sketched above, i.e., that a significant fraction of the luminosity of a nonthermal galactic nucleus is by way of very energetic particles. The energy density in the universe due to emission by Seyfert and QSOs has been conservatively estimated at 10^{-3} eV/cm^3. If Seyferts do indeed emit $\sim 10\%$ of their luminosity in the form of high-energy neutrinos, the energy density in such neutrinos could be as high as 10^{-4} eV cm^{-3}. Thus, one could reasonably hope to discover a cosmic background of high-energy neutrinos standing out above the atmosphere at $E_\nu > 10^{12}$ eV.

SUMMARY

Many of the most exciting objects in the sky may be high-energy neutrino emitters. Any object that emits gammas via π^0 production will also naturally be a neutrino source. Thus gamma-ray astronomy gives us some degree of confidence that neutrino astronomy will not be a null field. However, the two fields are complementary and will not merely duplicate each other, because neutrinos are capable of being seen even when they are produced in regions which are optically thick to gamma rays. For example, the very compact central regions of galaxies, with their intense radiation fields, will be optically thick to gamma rays but can be studied by neutrino astronomy. In probing these very central regions, one may find the details of how these objects are powered. In fact, in this case, even a null result is interesting because it would indicate the degree to which high-energy particles are involved.

Whenever astronomers have opened a new channel for studying the universe, new and unexpected sources have been discovered. One merely has to remember quasars and pulsars from radio astronomy, Cygnus X-1 from x rays, and the still mysterious x-ray and gamma-ray Bursters, to realize that the most exciting possible discoveries from neutrino-astronomy may not have been mentioned in this talk.

I am indebted to David Schramm, Fred Reines, Art Roberts, Maury Shapiro, Rein Silberberg, Steve Margolis, Ken Lande, John Learned, Ben Berezinsky, and John Scott, and many others for stimulating discussions and suggestions relating to the topic of extraterrestrial neutrinos.

REFERENCES

1. See the articles of D. Cline, J. G. Learned, and L. R. Sulak in these proceedings.
2. D. Eichler, Astrophys. J. (1978) (in press).
3. R. C. Lamb, Nature 272, 429 (1978).
4. V. S. Berezinsky, Proceedings of the International Neutrino Conference, Elbrus, U.S.S.R., 1977 (unpublished).
5. D. Eichler, submitted to Nature (1978).
6. D. J. Helfand, and E. Tademaru, Astrophys. J. 216, 842 (1977).
7. A. A. Stepanian, B. M. Vladimirsky, and Yu. I. Neshpot, Proceedings of the Fifteenth International Conference on Cosmic Rays, Plovdiv, 1977 (Bulgarian Academy of Sciences, Plovdiv, Bulgaria, 1977), Vol. 1, p. 135.
8. S. A. Colgate, and M. H. Johnson, Phys. Rev. Lett. 5, 235 (1960).
9. D. Eichler, submitted to Astrophys. J. (1978).
10. J. E. Grindlay, H. F. Helmken, R. Hanbury Brown, J. David, L. R. Allen, Astrophys. J. Lett. 197, L9 (1975).
11. N. A. Christiansen, D. Eichler, A P. Marscher, J. S. Scott, and W. T. Vestrand, Neutrinos-78, proceedings of the International Conference on Neutrino Physics and Neutrino Astrophysics, West Lafayette, Indiana, 1978, edited by E. C. Fowler (Purdue University, 1978), p. C7.

THE STUDY OF ULTRAHIGH-ENERGY NEUTRINO INTERACTIONS IN DUMAND

David Cline

Fermilab, Batavia, Illinois 60510

and

University of Wisconsin, Madison, Wisconsin 53706

ABSTRACT

The study of high-energy neutrino interactions in a large water-Čerenkov detector (10^9 tons) is shown to be feasible despite the limited energy resolution of such a device. The effects of an intermediate vector boson of mass less than ~70 GeV may be observable. The implications of a copious source of directly produced neutrinos are explored. In particular, this may provide a mechanism for the direct observation of the charged intermediate vector boson and the study of the "natural" neutrino flux at ~200 TeV.

INTRODUCTION

The study of high-energy neutrino interactions has become a focal point of modern elementary-particle physics ever since the discovery of weak neutral currents at CERN and Fermilab.[1] Thus, neutrino physics has already proven to be rich with new physics. If we extrapolate to high energies, it is easy to imagine new phenomena including W production, new leptons and Higgs boson production, and so forth. Can a cosmic-ray neutrino detector be devised to study these processes? This is one of the central issues in the design of the DUMAND array.[2,3]

The neutrino energy range of existing or soon to be completed accelerators is $\lesssim 1$ TeV. We anticipate that this energy range will be well explored over the next decade. In about a decade, the next generation of accelerators may be started. It is unlikely that the neutrino energy range will be extended beyond ~5-10 TeV. Other colliding beam lepton-hadron interaction machines may also be constructed, giving an equivalent energy of ~few TeV, but with modest luminosity compared to accelerator neutrino beams. Thus, for the next two decades it seems unlikely that terrestial neutrino sources or colliding lepton-hadron machines will exceed an equivalent laboratory energy of 5-10 TeV. In contrast, there are cosmic neutrino sources that extend up to 10^8 TeV and with appreciable neutrino flux above 10 TeV. If it can be demonstrated that a detector can be constructed to carry out neutrino physics with these neutrinos, it would appear that an exciting complementary source of information about elementary particles may exist. It is likely that proton-proton or proton-antiproton colliding beam machines will reach the equivalent laboratory energy of ~300 TeV within the next few years.[4] Thus, the existence of the intermediate vector boson may be resolved before a detector like DUMAND can be implemented.[5] Nevertheless, the higher-energy neutrinos carry astrophysical information as well as providing a source for neutrino physics, and this astrophysical information is possibly unique. Thus, a high-energy neutrino detector is also, in a sense, a neutrino telescope.[6] Furthermore, the behavior of the weak interaction at energies well beyond the W threshold will be of great interest and can only be settled by a

ISSN:0094-243X/79/520043-015$1.50

lepton-hadron scattering experiment that is directly sensitive to weak interactions. In particular, the neutrino cross section at energies in excess of 10^{13} eV will be of interest.

Before the detailed design of a detector to study high-energy neutrino collisions can be undertaken, it is essential to show that neutrino physics can be carried out with cosmic neutrinos; in contrast to neutrino event observation. All present cosmic-ray neutrino experiments are information-limited and simply record the existence of an event and some crude information about the direction of the muon that presumably comes from the neutrino collisions. If neutrino physics is to be carried out at such high energies, the detector will almost certainly roughly follow the design of the calorimeter-target neutrino detectors at accelerators such as the Harvard-Penn-Wisconsin-Fermilab or Cal Tech-Fermilab detectors at Fermilab.

ULTRA HIGH ENERGY NEUTRINO PHYSICS

The various Feynman diagrams that presumably are important in high-energy neutrino collisions are shown in Figs. 1, 2, 3, together with a DUMAND event diagram. There are two important assumptions:

1. The quark-parton model continues to be a useful description of neutrino-hadron interactions in the multi-TeV energy range.

2. There exists an intermediate vector boson that mediates weak interactions.

Two scaling variables seem particularly appropriate to the description of the scattering process:

$$x = Q^2/2m_p E_H \cong \frac{E_H E_\mu}{(E_\mu + E_H)m_p}[1 - \cos(\theta_\mu + \theta_H)],$$

$$y = E_H/(E_H + E_\mu),$$

where the symbols used here and below are defined by (see Fig. 1):

E_ν = neutrino energy
E_μ, θ_μ = muon energy and angle between the muon and neutrino 3-momenta
E_H, θ_H = energy of hadron jet, and angle between the hadron jet and neutrino 3-momenta
Q = difference between the neutrino and muon 4-momenta
m_p = proton mass; further:
M_w = mass of the intermediate boson
G = $(1.02 \times 10^{-5}\ m_p^{-2})$ = universal weak Fermi coupling constant.

Using these variables and definitions, we can write the cross section for the scattering on isoscalar targets as

$$\frac{d\sigma^{\nu,\bar{\nu}}}{dxdy} = \frac{G^2 m_p E_\nu}{\pi}\left[\frac{M_w^2}{M_w^2 + Q^2}\right]^2$$

$$\times\left[(1 - y + \frac{1}{2}y^2)\,F_2 - \frac{1}{2}y^2 F_L \mp y(1 - \frac{1}{2}y)xF_3\right],$$

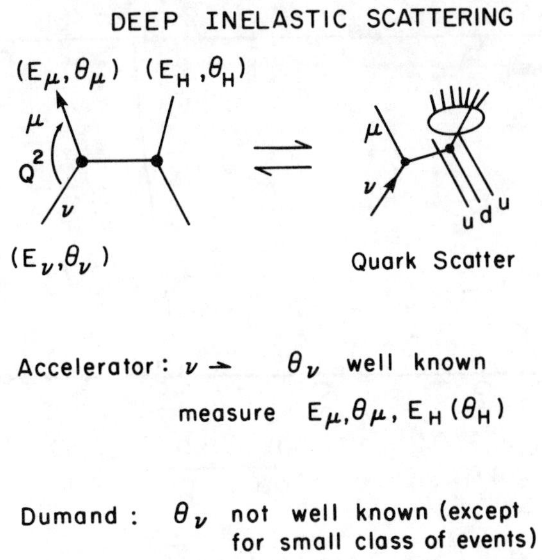

DEEP INELASTIC SCATTERING

(E_μ, θ_μ) (E_H, θ_H)

μ

Q^2

ν

(E_ν, θ_ν)

μ

ν

u d u

Quark Scatter

Accelerator : $\nu \rightarrow$ θ_ν well known

measure $E_\mu, \theta_\mu, E_H\ (\theta_H)$

Dumand : θ_ν not well known (except for small class of events)

Fig. 1. Variables in neutrino interactions.

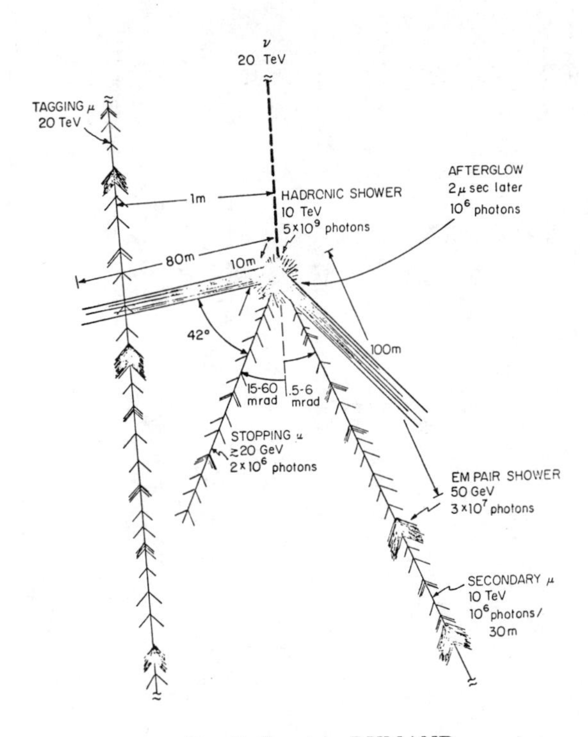

Fig. 2. Event in DUMAND.

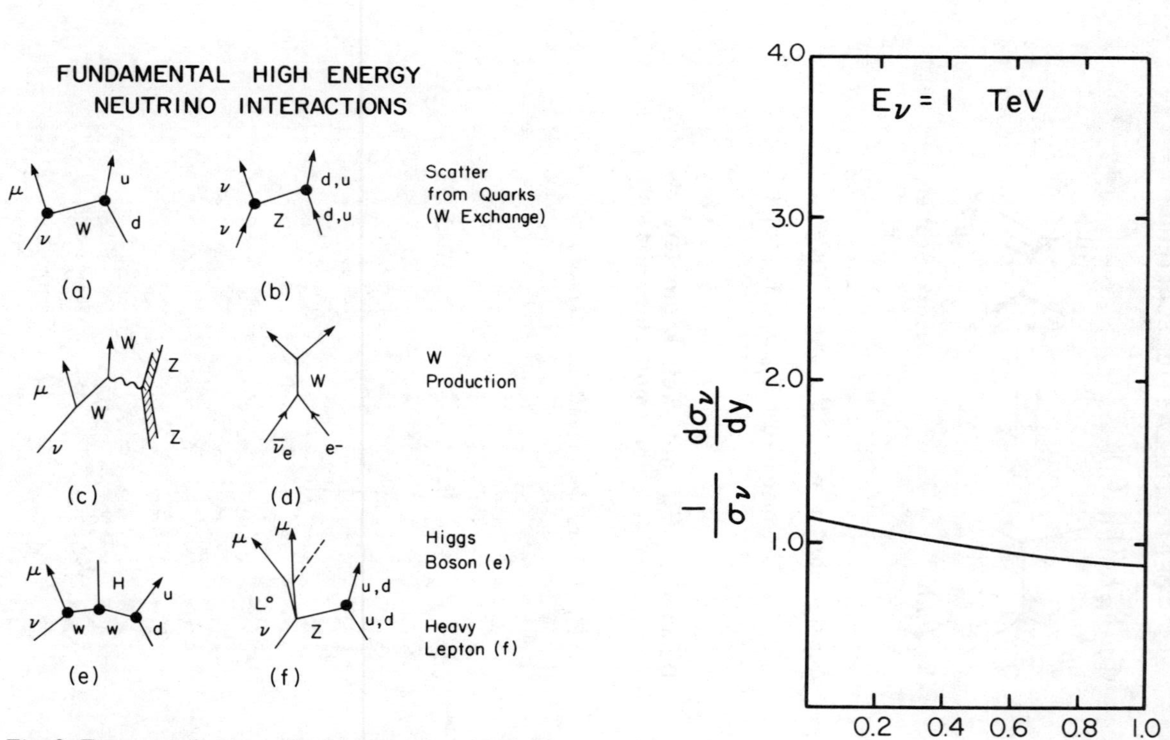

Fig. 3. Feynman diagrams for high-energy neutrino collisions.

Fig. 4. y distribution at 1 TeV in Weinberg—Salam model.

F_2, F_L, F_3: structure functions,
or in the quark-parton model,

$$\frac{d\sigma^\nu}{dxdy} = \frac{G^2 m_p E_\nu}{\pi}\left[\frac{M_w^2}{M_w^2 + Q^2}\right]^2 [xd(x) + x\bar{u}(x)(1-y)^2],$$

where d(x) and $\bar{u}(x)$ denote quark and antiquark x distributions, respectively. From present data, we find $x\bar{u} \ll xd(x)$ and thus the y dependence comes primarily from the term

$$\frac{M_w^2}{M_w^2 + Q^2} = \frac{1}{1 + xy\left[\frac{2m_p E_\nu}{M_w^2}\right]} \quad .$$

This term gives a clear signature for W at ultrahigh energy, i.e.,

$$2mE_\nu \ll M_w^2 \text{ and } xy\,\frac{2mE_\nu}{M_w^2} \gg 1.$$

In this limit, we find:

$$\frac{d\sigma^\nu}{dxdy} \to \frac{G^2 m_p E_\nu}{\pi}\left[\frac{M_w^2}{xy(2m_p E_\nu)}\right]^2 [xd + x\bar{u}(1-y^2)]$$

$$\to \frac{G^2 M_w^4}{4\pi M_p E_\nu}\left[\frac{xd(x)}{y^2}\right]\frac{1}{x^2},$$

$$\frac{d\sigma}{dy} \cong \frac{1}{y^2}\,\frac{G^2 M_w^4}{4\pi E_\nu m_p}\int_0^1 \frac{[xd(x)]}{x^2}\,dx.$$

Note two important properties of these distributions:

(i) The y distribution is characteristic for the W exchange model.

(ii) The collisions become more "elastic" with increasing energy ($E_\mu \to E_\nu$ ($E_H/E_\nu \to 0$)).

This is illustrated in Figs. 4, 5, and 6, where distributions at different energies are shown, together with a typical event at 100 TeV. The x distribution also becomes very sharp, i.e.,

$$\frac{d\sigma^\nu}{dxdy} \sim \frac{(xd(x))}{x^2} \sim \frac{F_2(x)}{x^2}, \qquad \langle x \rangle \to 0 \text{ as } E_\nu \to \infty .$$

Fig. 5. y-distribution at 100 TeV in Weinberg—Salam model and other possible processes.

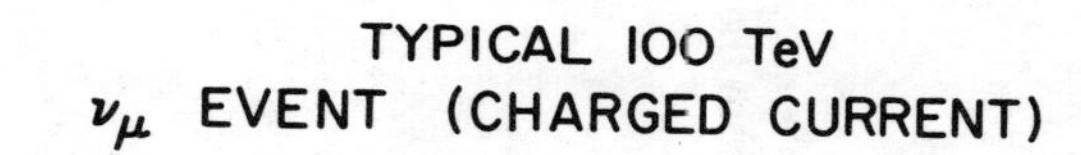

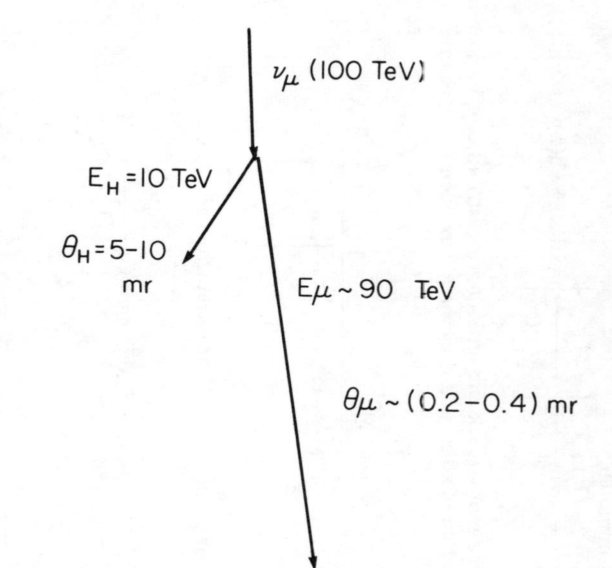

Fig. 6. Typical energy and angular division in high-energy collisions.

Note that we can write

$$\theta_\mu \cong \sqrt{\frac{2m_p}{E_\nu}\frac{xy}{1-y}} \rightarrow 0 \text{ as x or y} \rightarrow 0,$$

$$\sin\theta_H = \sqrt{\frac{E_\mu}{E_\nu}\frac{\cos^2\frac{1}{2}\theta_\mu}{1+E_H^2/Q^2}} \sim \cos\frac{1}{2}\theta_\mu \rightarrow 1 \text{ as } E_\mu \rightarrow E_\nu,$$

$$(y \rightarrow 0 \text{ and } E_H^2/Q^2 \rightarrow 0)$$

Therefore, the μ tends to follow exactly the neutrino direction, but the hadronic jet comes out at angles $\theta_H \gg \theta_\mu$ in the limiting cases indicated.

In summary, the expected event characteristics if an intermediate vector boson exists with $M_w \sim 70$ GeV, and if the Weinberg-Salam model is assumed, are:

(i) $\frac{1}{\sigma}\frac{d\sigma}{dy} \sim 1/y^2 \rightarrow$ events very elastic, $E_\mu \sim E_\nu$, $\theta_\mu \sim 0$
$\langle y \rangle \rightarrow 0$ (Fig. 4, 5, 6),

(ii) $\frac{1}{\sigma}\frac{d\sigma}{dx} \sim \frac{F_2(x)}{x^2} \rightarrow$ events almost "quasielastic",
$\langle x \rangle \rightarrow 0$, $\theta_H \rightarrow \gg \theta_\mu$, $E_H/E_\nu \rightarrow 0$,

and the total cross section rises slowly with energy:

(iii) $\sigma \rightarrow (10^{-34} \ln E_\nu)\ \text{cm}^{-2}$.

EVENT SIGNATURE AND DETECTOR RESOLUTION

To study neutrino interactions, it is essential to first prove that a neutrino event occurred in the detector and, second, to obtain kinematic information about the event. The latter may be easier than the former since in DUMAND there will be many thousands of times more muons incident on the detector than neutrino interactions. Furthermore, multi-TeV muons can make spectacular hadronic interactions similar to neutrino collisions. At first sight, it would seem possible to simply take neutrino events coming up from the earth but the rate is reduced and the earth may be opaque at very high energy. Note that accelerator experiments require a great deal of information in order to correctly separate a neutrino event and muon event; the most useful information on this point is due to veto detectors. Nevertheless, even with this feature, the selection of a neutrino event could be ambiguous and a detailed background study is necessary.

Assuming that neutrino events are separated from background, what can we learn about the collision in such a large detector? The first quantity of interest is the neutrino energy. Typical transition curves for Čerenkov light in water are shown in Fig. 7. The shapes of the curves are very insensitive to the total energy. Thus, the total light output is simply related to the hadronic energy. Using the Čerenkov light from the two components of a charged current interaction, it appears possible in principle to obtain the separate hadron and muon energies. However, the precision will necessarily be limited because of the limited sampling of the hadronic or electronic shower and the method of measuring the muon energy. My guess is that the neutrino energy can be obtained to <u>no</u>

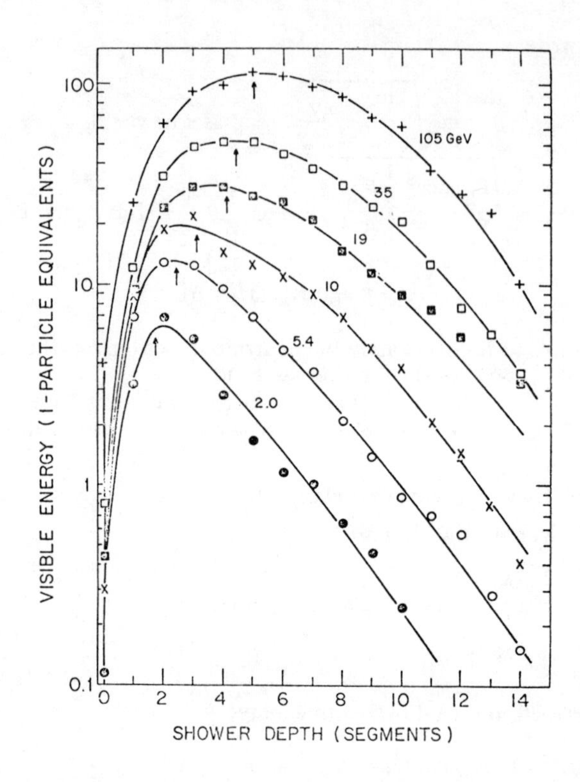

Fig. 7. Hadronic cascade shower development in the HPWF detector at Fermilab.

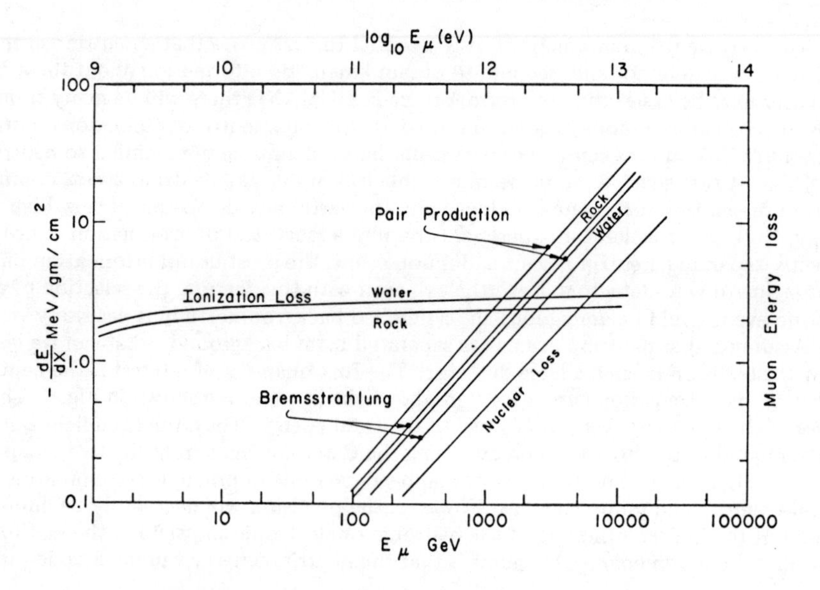

Fig. 8. Sources of muonic energy loss vs. energy.

better than a factor of two and it could be appreciably worse given the rapid fall-off of the neutrino spectrum. We consider that a charged current neutrino event is separated into four variables: the hadronic energy E_H, hadronic jet direction, muonic energy E_μ, and muonic direction.

The energy of the muon from the neutrino collision can be obtained by measuring the electromagnetic energy and hadronic collision energy loss. This technique will be useful in the energy range above ~1 TeV, since then these processes become appreciable in cross section. Fig. 8 shows the relative energy loss by the various mechanisms for high-energy muons. The direction of the muon can be obtained from pulse height and timing information of the Čerenkov light pulse. The DUMAND neutrino-signature group has considered this in more detail, and has concluded that it is feasible.

The hadronic "jet" from the neutrino collision can also be observed. The total-energy measurement can be obtained from the pulse height of Čerenkov light once the position of the vertex is known. The vertex can be determined using the $\pi \to \mu \to e$ decay processes left from the hadronic cascade. Once the vertex is known, the pulse heights can be "corrected" and an estimate of the total energy obtained. To reduce the uncertainty in this technique, one can use the muon-induced nuclear interactions, which can be separated from the bremsstrahlung by the residual $\pi \to \mu \to e$ decay. The properties of inelastic N scattering are well known and the muon spectrum is measured principally from the electromagnetic burst spectrum with a large number of events. It should be possible to obtain a calibration of the detector in this way.

The measurement of the hadronic jet direction is considerably more difficult, since the "jet" cascade terminates rapidly in the water and thus the lever arm is small. Furthermore, it is unlikely that more than a few photo-multipliers will "see" the photons from the jet. Finally, the Čerenkov light angle is 41° and is smeared out by the width of the cascade due to the hadronic processes. Hence, the measurement of the jet direction will be less accurate.

To summarize, E_μ and E_H can be measured to within a factor of 2 or possibly better; the muonic direction can be well measured if E_μ is very large so that the muon traverses the full detector; the hadronic jet direction can only be poorly measured. For events with an accompanying partner μ (from π or K decay in the atmosphere), it will be possible to estimate the incident neutrino direction; otherwise this must be obtained from the measurements of E_μ, E_H, and the corresponding directions. In Table I, a rough estimate is given of the expected resolution of these parameters.

Table I. Estimated angular and energy resolution in DUMAND

Quantity	Resolution	Assumptions
H-jet direction	$\sim \frac{10\ \text{cm}}{120\ \text{m}} \sim 1$ mrad	4 photomultipliers hit; 5 photoelectrons obtained in each; timing resolution $\sigma_t \sim 0.5$ns
μ direction	$\sim \frac{10\ \text{cm}}{500\ \text{m}} \sim 0.2$ mrad	muon passes through half of detector; $\sigma_t \sim 0.5$ns
E_H	~ (30-50)%	event vertex can be determined; 4 p.m. hit, 5 photoelectrons obtained in each
E_μ	~ 20%	100 measurements
	~ 50%	40 measurements

Note that in obtaining the neutrino direction to fix the origin of the neutrino flux, knowledge of E_H and the muonic direction allows a good estimate of the neutrino direction on average. This will be useful for neutrino astronomy. The resolution of the detector for the x and y scaling variables can be estimated as follows in the small angle approximation:

$$x = \frac{E_\mu E_H (\theta_\mu + \theta_H)^2}{2m_p (E_H + E_\mu)}, \quad y = E_H/(E_H + E_\mu),$$

and the resulting resolution functions are

$$\left(\frac{\delta x}{x}\right)^2 = 4\frac{[\delta(\theta_\mu + \theta_H)]^2}{(\theta_\mu + \theta_H)^2} + \left(y\frac{\delta E_\mu}{E_\mu}\right)^2 + (1-y)^2\left(\frac{\delta E_H}{E_H}\right)^2,$$

$$\left(\frac{\delta y}{y}\right)^2 = (1-y)^2\left(\frac{\delta E_\mu}{E_\mu}\right)^2 + (1-y)^2\left(\frac{\delta E_H}{E_H}\right)^2.$$

Consider the following typical examples:

(i) E_ν = 100 TeV, E_H = 10 TeV, E_μ = 90 TeV, $\delta(\theta_\mu + \theta_H)/(\theta_\mu + \theta_H) = \frac{1}{4}$,
$\delta E_\mu/E_\mu = \delta E_H/E_H = \frac{1}{2}$, x = 0.1, y= 0.1,
$\delta x = 0.07$, $\delta y = 0.06$. Rough estimates of x,y are possible even with a 40-m spacing between the photomultipliers of the DUMAND array.

(ii) For E_ν = 100 TeV, y = 0.9, x = 0.1,
$\delta x = 0.07$, $\delta y = 0.06$.

Thus we find that event by event the x and y can be determined to the accuracy required to distinguish the existence of events with large y and, thus, test for the expected rapid change of y distribution shown in Figs. 4 and 5.

NEUTRINO FLUX ESTIMATES

(i) Pion and Kaon Decays in the Atmosphere

Neutrinos are ultimately produced in any collision where π and K mesons and μ fermions are produced through decays. These collisions may occur in the atmosphere, in the disk of a supernova, in the metagalaxy, or in the original big bang. In addition, there may exist several sources of prompt neutrinos through the production and decay of particles with very short lifetimes. One example would be charmed-particle production and decay. For example, the production of prompt neutrinos is implied by the data on the neutrino production of charged dilepton pairs. The production of other short-lived particles, such as unstable quarks or charged and neutral intermediate vector bosons, could lead to new sources of neutrinos. For neutrinos produced in the atmosphere, it is expected that any novel source will produce prompt muons as well. At the higher energies, the neutrino spectrum can be inferred from the muon spectrum; this procedure is valid only if the π/K ratio is known and if the muons originate from either π or K decay. For example, the electromagnetic pair production of μs would not contribute appreciably to the neutrino spectrum. The probability of π or K decay rapidly decreases with energy and the calculated neutrino flux up to 100 TeV is shown in Fig. 9. Above 100 TeV, the flux is uncertain for the reasons stated above. Nevertheless, we may guess that certain energy intervals will contain information on specific sources. For example, above $E_{\nu} > 10^{15}$ eV it is unlikely that atmospheric π and K decay neutrinos can be detected even in DUMAND but prompt neutrinos might be observed (see below).

In this same energy region, the neutrinos from metagalactic sources may become significant according to the estimates of Berezinsky et al.[3] Unfortunately, these neutrinos are not accompanied by muons or electrons (the muons decayed long before and the electrons were degraded by starlight) and thus there is no direct experimental estimate of the neutrino flux. The flux of high-energy photons does, however, provide an upper limit on the neutrino flux.

One procedure to "measure" the ν spectrum is to infer it from the cosmic μ spectrum (latest data in Fig. 10; measurements of the μ flux in DUMAND will be crucial for this flux measurement).

(ii) Prompt Neutrinos

The production of charmed particles and subsequent semileptonic decay will lead to prompt neutrinos produced within $\sim 10^{-13}$ sec after the initial interaction. In contrast to π-decay and K-decay neutrinos, these neutrinos will not be attenuated by nuclear absorption of secondaries and thus the spectrum will fall off by one power of E_{ν} less than the π and K neutrinos. Thus, at some energy this spectrum will dominate the other atmospheric sources.

It is difficult to estimate the cross section for charm production, but recent beam-dump experiments at CERN indicate that the cross section may be in the vicinity of $\sim 100\ \mu b$ for 400 GeV incident protons.[6] The cross section may reach even higher values at ultrahigh energies.

We have a poor understanding of the origin of the prompt neutrinos observed at CERN. However, if they arise from charm pair production it seems reasonable that there will be a new source of neutrinos for use in DUMAND. We estimate that the prompt neutrinos will dominate the atmospheric π-decay and K-decay neutrinos above 50 TeV.

An important feature of the new direct neutrino results is that the fraction of $\bar{\nu}_e$ produced is measured to be $\sim 1/4$ while $\bar{\nu}_e$ are strongly suppressed in the atmospheric fluxes of decay neutrinos due to the long muon lifetime and the other mode being via rare K decay. Thus, not only will direct neutrinos dominate the ultra-high energy cosmic ray neutrinos, but there will be large numbers of $\bar{\nu}_e$, which will enable production of the W through resonant reactions $\bar{\nu}_e + e^- \rightarrow W^- \rightarrow$ all. It is clearly very important to study the prompt ν production at accelerators.

POSSIBLE EXPERIMENTS IN THREE EXAMPLES

Using the estimated detector resolution, the estimated neutrino spectrum (including the prompt neutrino rate) and the expected behavior of neutrino interactions in the Weinberg-Salam model, we can guess some of the experimental results which may be obtained in the first two years of DUMAND operation. We emphasize that these results are only suggestive. Detailed calculations are needed.

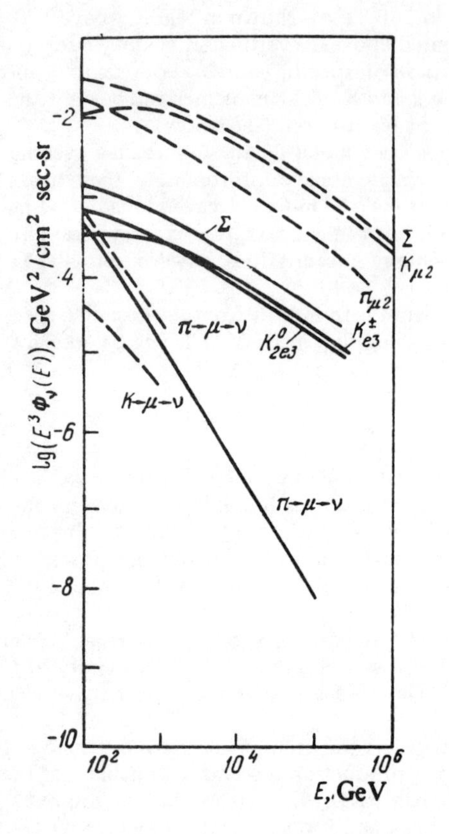

Fig. 9 Vertical atmospheric neutrino flux estimated in Ref. 3.

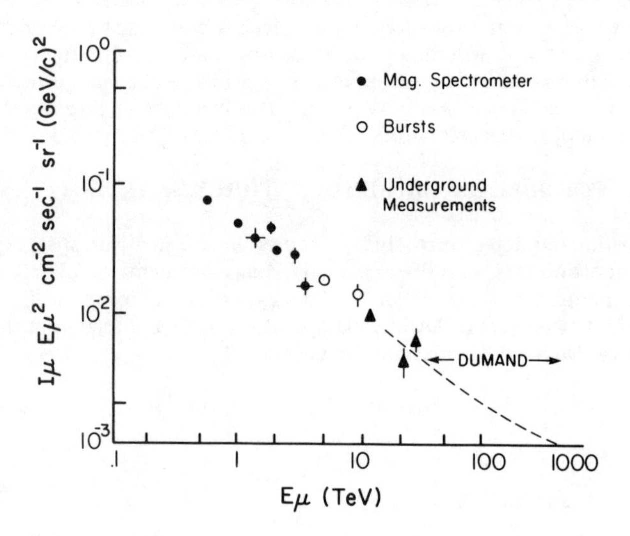

Fig. 10. Measured cosmic muon spectrum at high energy.

(i) Measurement of the Mean y

The quantity $\langle y \rangle = \langle E_H \rangle / \langle E_H + E_\mu \rangle$, when measured as a function of $E_{visible} \sim E_H + E_\mu$, can give important information about the structure of weak interactions at high energy, as shown in Fig. 11. This quantity is also relatively insensitive to the inherent experimental errors in the poor DUMAND resolution functions. A full Monte Carlo analysis is needed to assess the reliability of this measurement. However, we can roughly estimate the $\langle y \rangle$ measurements, as shown in Fig. 11. At high energies, it appears that $\langle y \rangle$ will be sensitive to the existence of the W boson.

(ii) Measurement of the Total Cross Section Above 10 TeV

A second, but much more difficult measurement that may be carried out in DUMAND is that of the total cross section. This requires neutrino flux measurements or reliable calculations. Using the observed muon spectrum above 50 TeV in DUMAND, it may be possible to reliably estimate the ν flux. This possibility needs considerable study. The expected behavior of the σ_{total} is shown in Fig. 12. After one year's operation, σ_{total} should be crudely known. If σ_{total} rises linearly with E_ν up to 100 TeV, the earth will become opaque and this would provide a flux-independent measure of σ_{total}.

(iii) Direct Excitation of the W^- by $\bar{\nu}_e$

Glashow[7] suggested the existence of the resonant reaction

$$\bar{\nu}_e + e^- \rightarrow W^- \rightarrow \bar{\nu}_\mu + \mu^- .$$

For very heavy W-bosons, a more effective reaction could be[8]

$$\bar{\nu}_e + e^- \rightarrow W^- \rightarrow \text{hadrons.}$$

In the Weinberg-Salam model, the resonant energy for the $\bar{\nu}_e + e^-$ reaction is (m_e being the electron mass):

$$E_0 = \frac{M_w^2}{2m_e} \approx 5 \times 10^{15} \text{ eV.}$$

Berezinsky, Cline, and Schramm[9] have estimated the production of the W^- using the prompt $\bar{\nu}_e$ neutrino sources discussed before.

The rate, η_{res}, of resonant events in a target with N_e electrons is

$$\eta_{res} = 2\pi\, N_e \sigma_{eff} \gamma \theta_{\bar{\nu}e} \quad (E_\nu > E_0),$$

for a power law spectrum with integral index γ and a 2π solid angle available for detection of the W-boson with $M_w \gtrsim 50$ GeV. The effective cross section is $\sigma_{eff} \approx (3\pi/\sqrt{2})G = 3 \times 10^{-32}$ cm^2 and for a 3×10^{11} cm^3 water detector, N_e is $\sim 10^{41}$. For an acoustic DUMAND array with $M \sim 10^{11}$ tons, the counting rate is ~20/year.

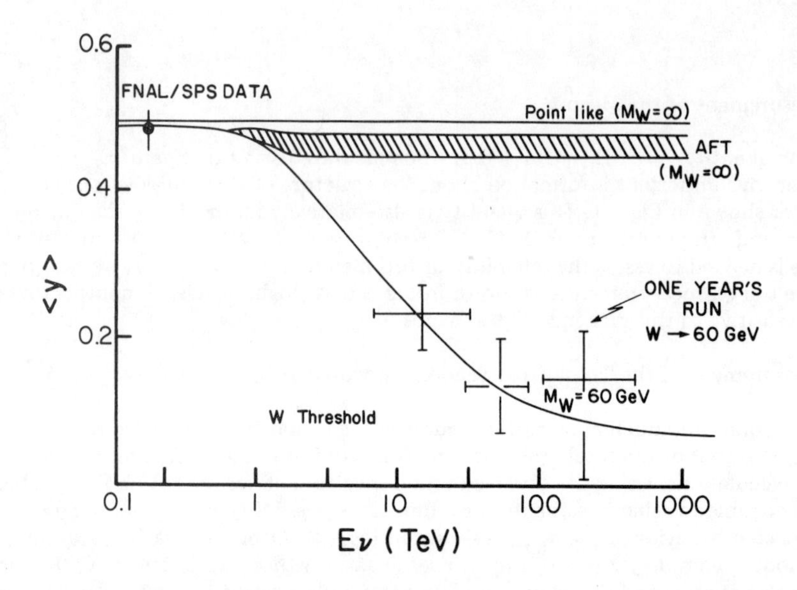

Fig. 11. Predicted mean y at high energy and hypothetical measurements in DUMAND in one year's run.

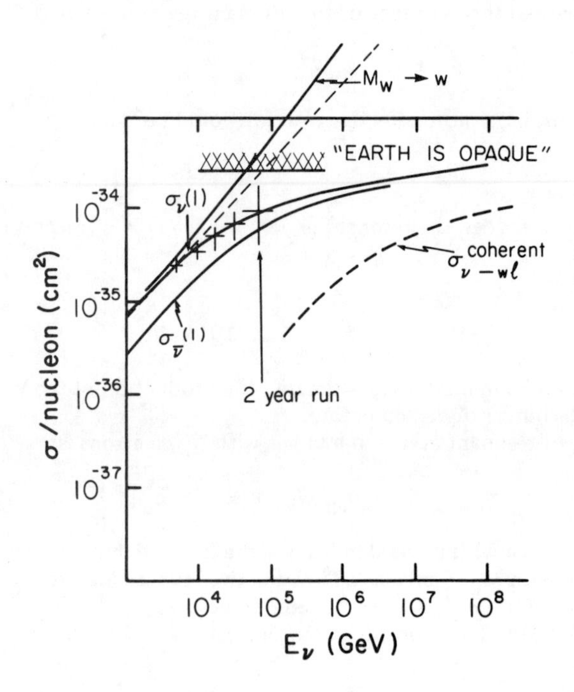

Fig. 12. Predicted total cross section σ and hypothetical measurement.

REFERENCES

1. G. Myatt, in Proceedings of the Sixth International Symposium on Electron and Photon Interactions at High Energies, Bonn, 1973, edited by H. Rollnik and W. Pfeil (North-Holland, Amsterdam, 1974), p. 389; F. J. Hasert et al., Phys. Lett. 46B, 138 (1973); A. Benvenuti et al., Phys. Rev. Lett. 32, 800 (1974); F. J. Hasert et al., Nucl. Phys. B73, 1 (1974).
2. Proceedings of 1975 DUMAND Summer Study, Bellingham, Washington, 1975, edited by P. Kotzer (Western Washington State College, Bellingham, Washington, 1976); Proceedings of 1976 DUMAND Summer Workshop, Honolulu, 1976, edited by A. Roberts (Fermilab, Batavia, Illinois, 1977); A. Roberts, Report on the 1977 DUMAND Summer Workshop, Moscow, 1977, Office of Naval Research (London Branch) Report No. ESN 31-9, 1977, p. 370 (unpublished).
3. V. S. Berezinsky and G. T. Zatsepin, Sov. Phys. Usp. 20, 361 (1977).
4. C. Rubbia, P. McIntyre, and D. Cline, in Neutrino '76, proceedings of the International Neutrino Conference, Aachen, 1976, edited by H. Faissner, H. Reithler, and P. Zerwas (Vieweg, Braunschweig, West Germany, 1977).
5. See also J. D. Bjorken, "Future Accelerators: Physics Issues," SLAC Report No. SLAC-PUB 2041, 1977 (unpublished), Fig. 1.
6. P. C. Bosetti et al., CERN preprint No. CERN/EP/Phys. 78-2, 1978; P. Albiran et al., CERN preprint No. CERN/EP/Phys. 78-3, 1978.
7. S. L. Glashow, Phys. Rev. 118, 316 (1960).
8. T. Gaisser and A. Halprin, Proceedings of the Fifteenth International Conference on Cosmic Rays, Plovdiv, Bulgaria, 1977 (Bulgarian Academy of Sciences, Plovdiv, Bulgaria, 1977), Vol. 6, p. 265.
9. V. Berezinsky, D. Cline, and D. Schramm, "Prompt Atmospheric Neutrino Production of W Bosons" (submitted to Phys. Lett., 1978).

PROJECT DUMAND AND THE TRADE—OFFS BETWEEN ACOUSTIC AND OPTICAL DETECTION

John G. Learned
Department of Physics, University of California, Irvine, California 92717

ABSTRACT

The Deep Underseas Muon and Neutrino Detection (DUMAND) Project is briefly described, with emphasis upon the trade-offs between optical and accoustic detection in the proposed cubic kilometer detector. The status of the program, which is nearing the detector design study and testing stage, is discussed.

INTRODUCTION

The Deep Underseas Muon and Neutrino Detection (DUMAND) Project has the twin goals of studying neutrino interactions at ultrahigh energies (> 1 TeV) and beginning the field of high-energy neutrino astronomy. The vanishingly small fluxes and minuscule weak interaction cross section force target volumes of 10^9 tons in order to achieve reasonable rates. The only medium affordable in this quantity (1 km^3) is the ocean, which serves as target, detection medium, and shield from near-surface backgrounds. The detection techniques available are twofold: light and sound originating from the cascade of particles ensuing from the neutrino-nucleon interaction. The Čerenkov light consists of an intense blue flash visible up to ~100 m from the interaction. The thermoacoustic pulse travels with little attenuation to kilometer distances with characteristic frequency in the 20 kHz region, but this pulse is intrinsically small and difficult to detect except at the highest energies. The technology for detection does exist, so that design is a matter of the interplay of physics, engineering, and economics.

The project is the creation of an organization that has been meeting regularly for the last five years, and has sixty members from twenty institutions in the United States and a number from abroad, particularly from the U.S.S.R. There have been three successive Summer Studies[1-3] with Proceedings, and a six-week long fourth workshop[4] was held at La Jolla in August 1978. A proposal for a two-year design study[5] has been submitted to U.S. funding agencies in March 1978. This proposal requests funds for a laboratory at The Scripps Institution in La Jolla, California, and site studies in Hawaii. The result of this effort will be largely a design, including critical component construction and testing, for a cubic kilometer detector.

The rest of this report will consist of an abbreviated summary of some physics aspects of the experiment followed by a discussion of the differences between acoustic and optical detection. For detailed discussions of ocean engineering, detector technology, site considerations, etc., the interested reader is referred to the Summer Studies[1-4] and the proposal.[5]

PHYSICS AND ASTRONOMY WITH DUMAND

A. Motivation for Experiment

As an introduction to discussing the sources of the neutrinos and their interactions, let us make a few observations upon the motivation for this project. Attendees of this conference will need little convincing of the importance of studying neutrino physics at energies beyond those currently available from accelerators.[6-7] By detection of the

ISSN:0094-243X/79/520058-27$1.50

hadronic cascade alone some high-energy physics results can be obtained,[8] but much more is gained from detection of any muons and measurement of their energies and angles. This now seems possible only with optical detectors, so when discussing a high-energy physics experiment we are automatically considering either a purely optical array or an array with both optical and acoustical ability. Fortunately, only poor energy resolution is required for reasonably good y resolution ($\sim$0.1),[9] because any DUMAND array of moderate cost will (obviously) be sparsely instrumented. The experiment, while not measuring charge, can thus measure the same sort of quantities that have been studied in the first generation of neutrino detectors at Fermilab, with much poorer resolution and rate,[10] but still with useful experiments possible.

On the astronomy/astrophysics side of the experiment, there is much activity among astrophysicists at present, considering possible neutrino sources and the implications of measurements of the fluxes. It is obvious that neutrinos provide a unique probe of the universe, arriving unattenuated from even dense sources, undeviated by magnetic fields or scattered by the intervening dust and starlight (gamma rays have a cut-off due to photon-photon scattering of $\sim 10^{14}$ eV for distances beyond the nearest galactic cluster, for one example). Neutrinos are able to arrive here even from the most distant sources and earliest times. With observation of the cascades alone, one achieves less than degree resolutions (getting increasingly better at higher energies), which is adequate for localizing point sources.

The key uncertainty in the speculations of what DUMAND can achieve is, of course, rate. The observed rate in the detector can be written as

$$R = \int_{E_{th}}^{E_{max}} \Phi_\nu(E)\sigma_{\nu N}(E)V_D(E)dE, \tag{1}$$

that is, as the integral of the product of flux (Φ_ν) times cross section ($\sigma_{\nu N}$) times effective target volume (V_D), from threshold (E_{th}) to maximum energy (E_{max}).

We will discuss the energy dependence of the effective volume later. For optical arrays, however, it will be essentially constant above a few hundred GeV. Since the acoustic case is more complex, we shall (conservatively) assume a constant mass of 10^9 tons for the remainder of this section. We will proceed to discuss neutrino sources, interaction cross sections and then rates, and the type of experiments that can be performed with DUMAND.

B. Neutrino Sources

We divide the high-energy neutrino sources into two classes, atmospheric and extraterrestrial, as indicated schematically in Fig. 1. The atmospheric neutrinos are the descendants of primary cosmic rays which strike the atmosphere, with progeny consisting mostly of pions and kaons. Due to the increasing laboratory lifetime with energy, fewer and fewer of these mesons decay prior to interaction, and the resulting neutrino spectrum is one power of the energy steeper than the primary spectrum (given certain other assumptions, e.g., scaling). These neutrinos are mostly muonic, since few muons have a chance to decay in flight and there are relatively few ($\sim$2%) from kaon decays. The flux of neutrinos is also stronger ($\sim$25%) in muon neutrinos than muon antineutrinos. Because of the difference between the y distributions for neutrinos and antineutrinos (constant versus $(1\text{-}y)^2$, for energies below a few TeV, anyway), the observed y distribution will depend upon this ratio.[11] Notice, as illustrated in Fig. 2, that all the neutral-current events and the electron neutrino events appear together as a spike at high y.

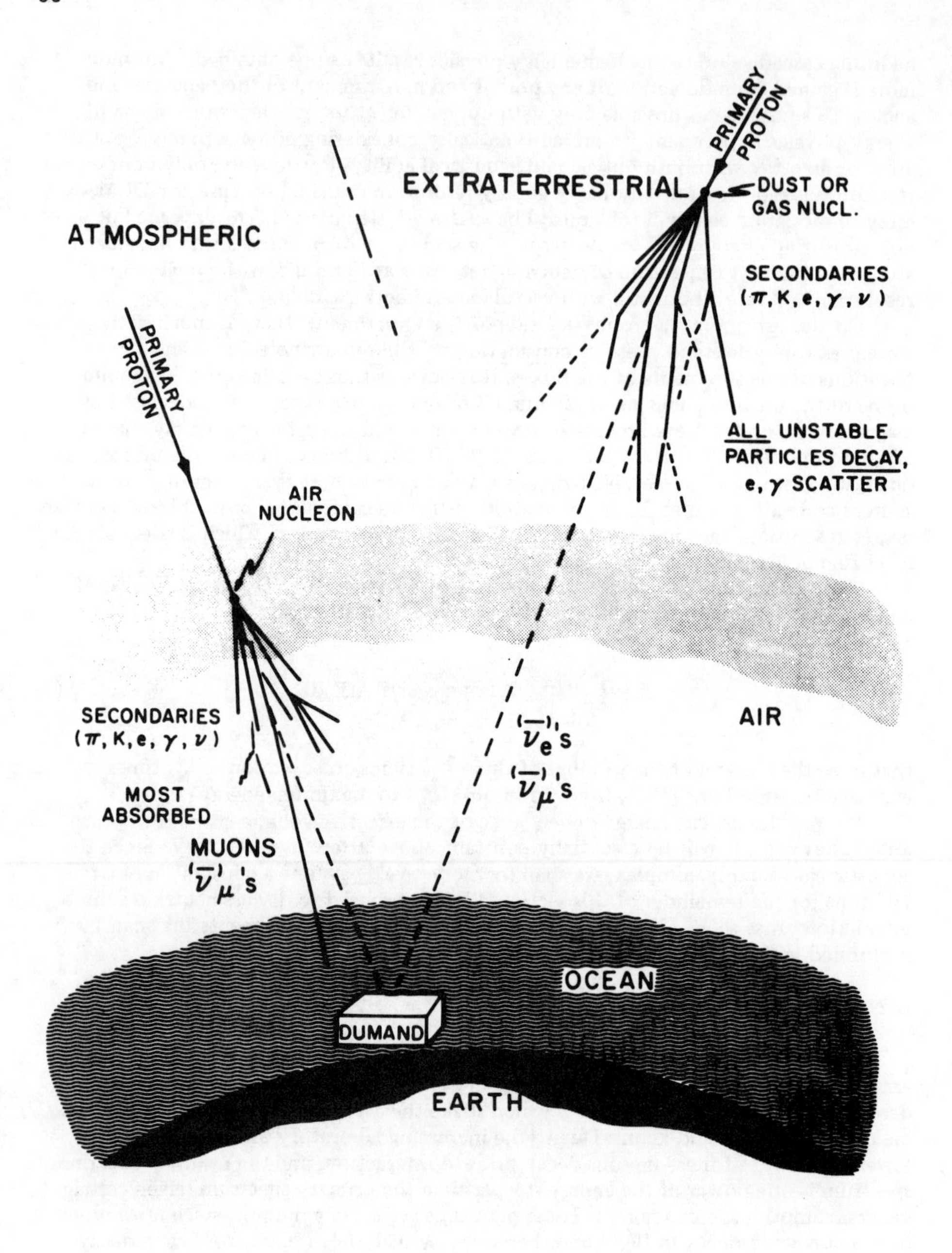

Fig. 1. Two classes of neutrino sources are indicated schematically.

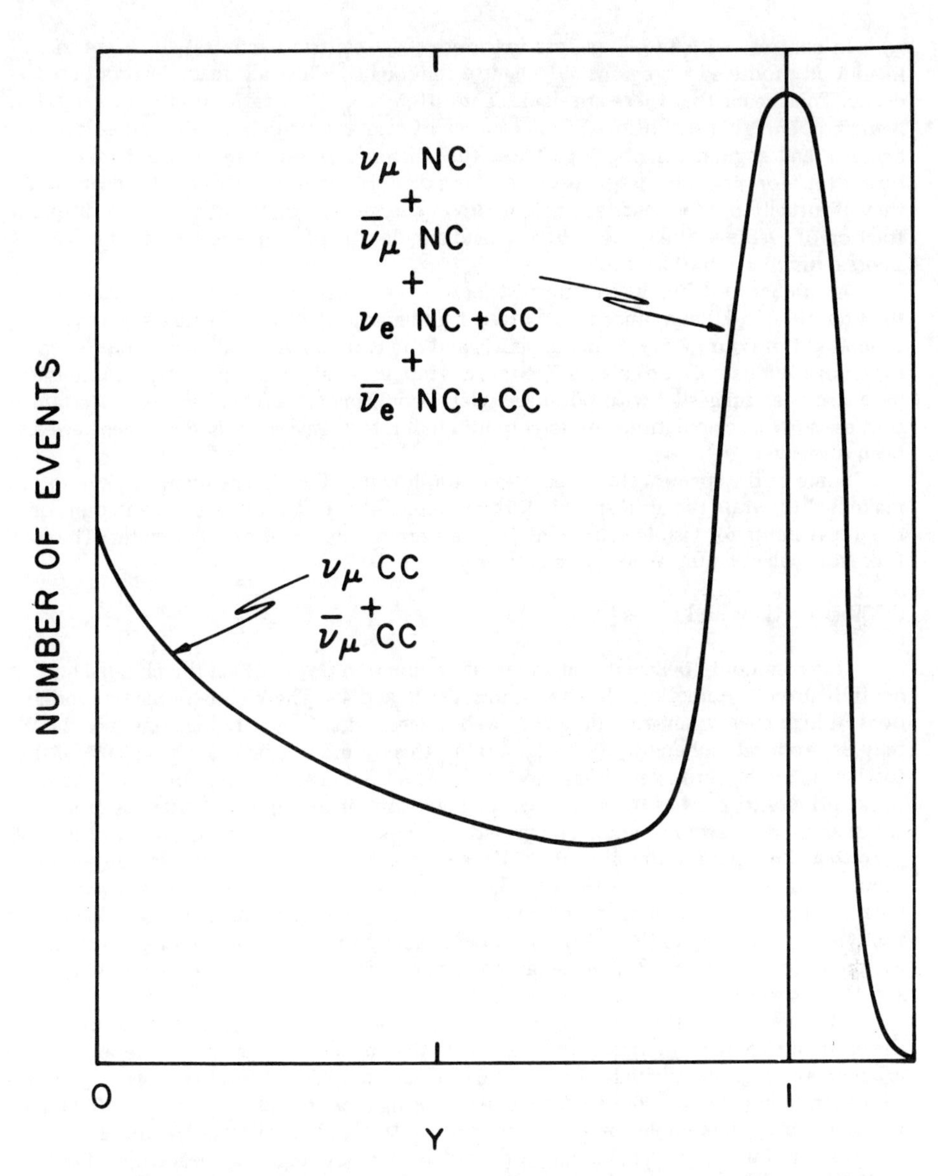

Fig. 2. Sketch of the observed y distribution which is the sum of electron and muon neutrino and antineutrino interactions, both charged and neutral current, showing effect of piling up of events at large y.

In contrast with the above are extraterrestrial neutrino sources which are in all likelihood produced in regions sufficiently tenuous to allow all unstable products to decay. This means that there are significant differences in the ratio of muon to electron neutrinos, energy distribution (given a source of primaries which has the same slope with energy), and angular distribution. These flux differences result in an event rate one power (or more) of the energy less steep for extraterrestrial sources (which means that they eventually predominate as one goes up in energy), a y distribution with a different fraction of events in the spike at high y, and angular distributions reflecting the stable sidereal distribution of the flux.

A number of different extraterrestrial sources of neutrinos have been discussed in the literature,[12-18] and more are at present under consideration. Diffuse sources are estimated from our galaxy, other galaxies, and the early universe. Point sources from supernova remnants, our galactic center, Seyfert galaxies, radio galaxies, and quasars have also been suggested with potentially large counting rates (or, given the uncertainties of these sorts of calculations, undetectably small rates). Several pulsed sources have also been suggested.

Some of the spectral flux predictions are shown in Fig. 3. The main observation to make is that, while the atmospheric flux represents a minimum flux one can count on, the upper limits for visible sources at 10^{15} eV are on the order of 10^6 greater. The total flux from galactic nucleii could be greater yet.

C. Cross Sections and Rates

It is commonly believed that the weak interaction proceeds via the charged and neutral-current interactions illustrated in Figs. 4a and 4b. The Weinberg-Salam model does fit high energy neutrino data very well indeed. At sufficiently high energies the W may be produced coherently (see Fig. 4c) but this should not be more than 10% of the total cross section until very high energies.[19] The total cross section does not level off, but continues to grow as the square of the logarithm of the energy (scattering from the sea quarks), as shown in Fig. 5. The growth in cross section occurs for events with small y, so that the distribution of events of large y (>0.5) comes to have the same spectral shape as the incoming flux (Fig. 6). The mean y value as a function of energy (for charged current interactions, averaged over neutrino type) is a simply measurable quantity that yields a clear separation of the Weinberg-Salam model from a cross section that continues to rise linearly.[6] This remains an important experiment, even if the W is observed at accelerators first.

Another effect of the rising cross section is the attenuation of neutrinos through the earth at very high energies. Observation of the angular distribution alone of events of given energy permits calculation of a total cross section. The effect becomes very strong for neutrino energies $\gtrsim 100$ TeV for a linearly rising cross section (down/up$\approx$20/1), and would be easily detectable for a one year's run with a 10^9 ton DUMAND using the atmospheric flux alone.[6] (This measurement would not require observation of the muon.)

Other interactions that might be studied include Higgs-boson production (Fig. 4e) proceeding via a Drell-Yan mechanism and, thus, having strong (E^2) energy dependence and the production of heavy leptons (Fig. 4f). The multimuon final states seen in accelerator studies can apparently be partially accounted for by charm and by muon bremsstrahlung, but it appears that a few unlikely events of extremely high energy and also (possibly) multi-strange-particle final states are hints of new phenomena at the 10^{-3} or 10^{-4} level. Since we do not know the mechanism(s) for these events, we can only speculate as to energy dependence. DUMAND would be sensitive to production of

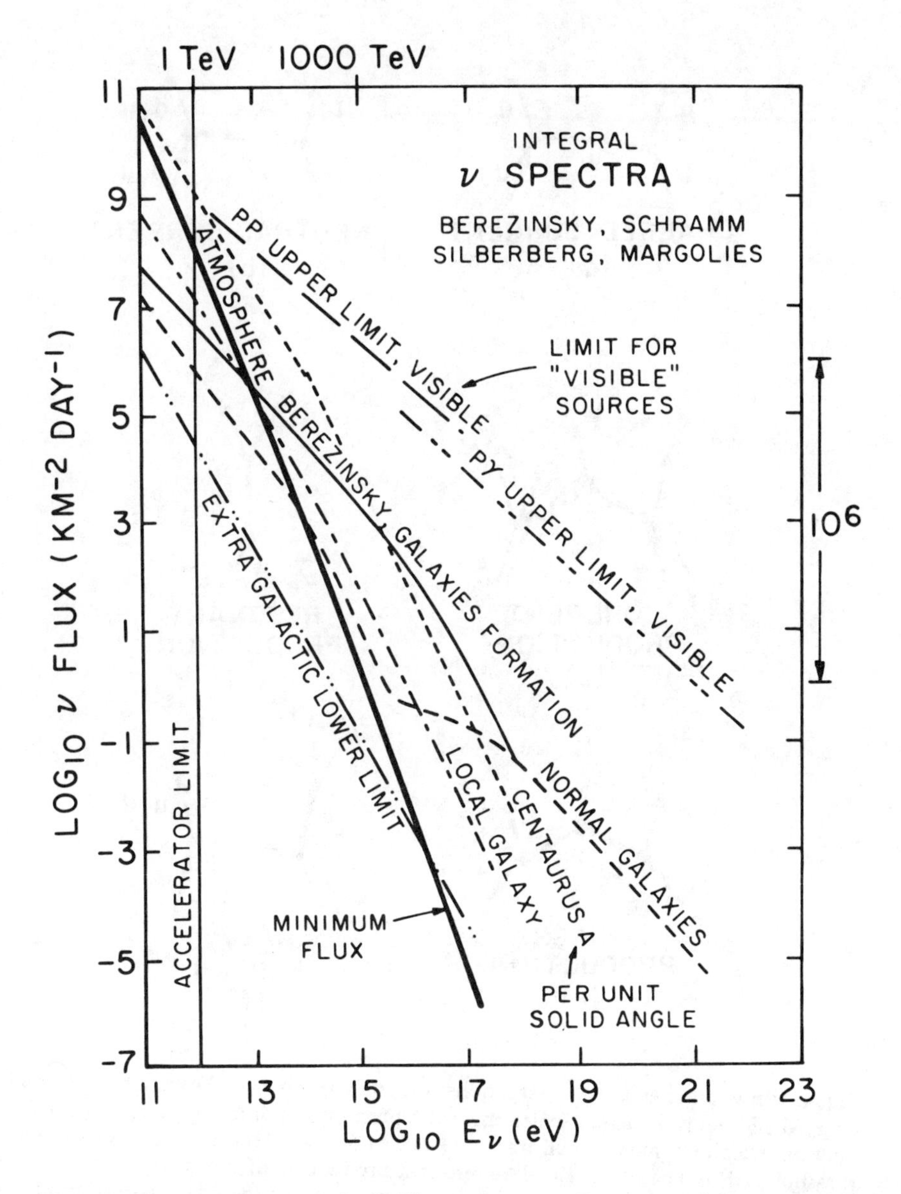

Fig. 3. Neutrino flux predictions.[12-18] Atmospheric flux extrapolation represents a minimum expected value. Upper limits[17] apply only to sources not heavily shielded. Some predictions[14] from galactic nuclei lie above these curves, but spectral distributions have not yet been predicted.

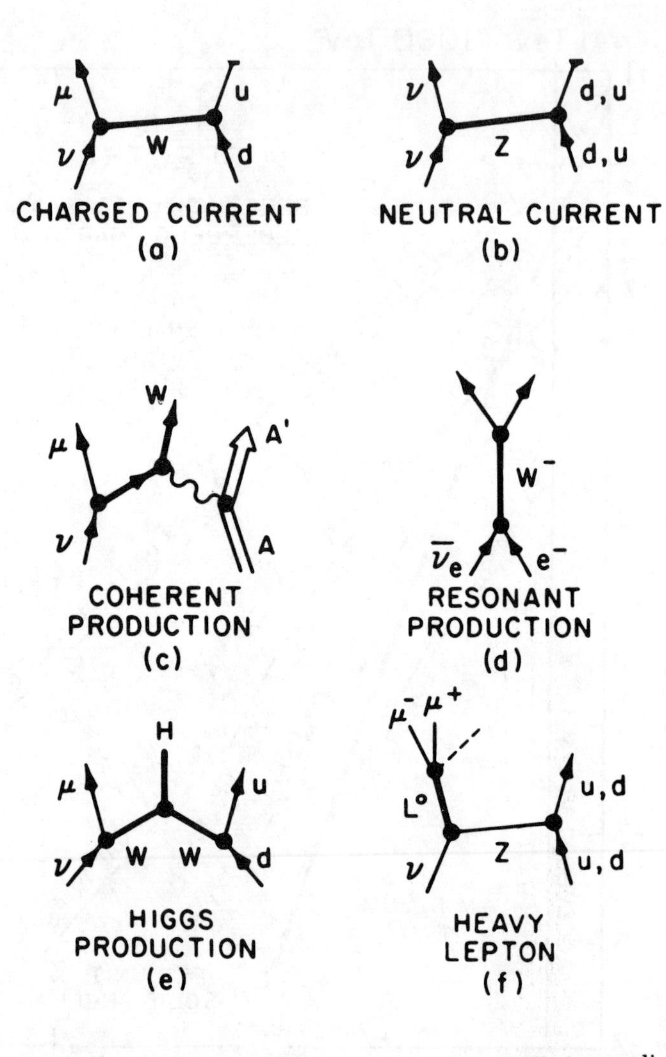

Fig. 4. Ultrahigh-energy ν-N interactions: (a) and (b) show the Feynman diagrams for charged and neutral-current scattering, (c) represents the dominant diagram for W production, which remains a small fraction (~10%) of the total cross section even above threshold; (d) indicates the timelike resonant production of the W^- (Glashow resonance). Higgs boson production (e), as well as other processes such as heavy lepton production (f), may also contribute significantly at extreme energies.

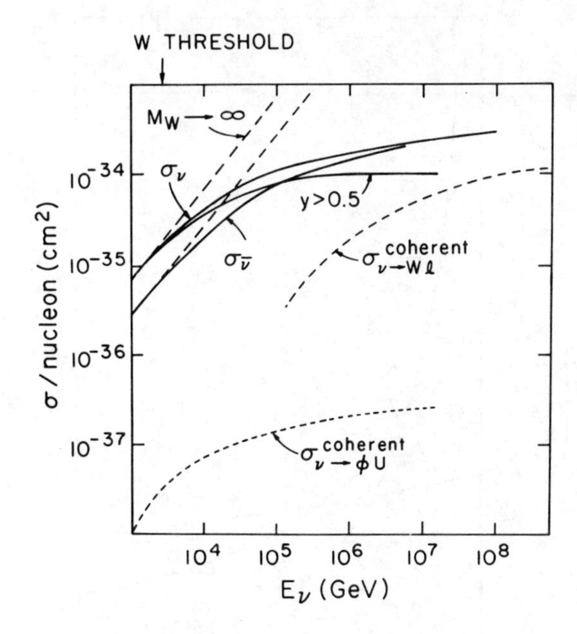

Fig. 5. Predicted neutrino and antineutrino-nucleon total cross sections above 1 TeV with an intermediate vector boson, as given in the Weinberg-Salam model. The cross sections continue to rise as $\ln^2 E_\nu$, but become asymptotically constant for large y (from Refs. 19 and 2 with modifications).

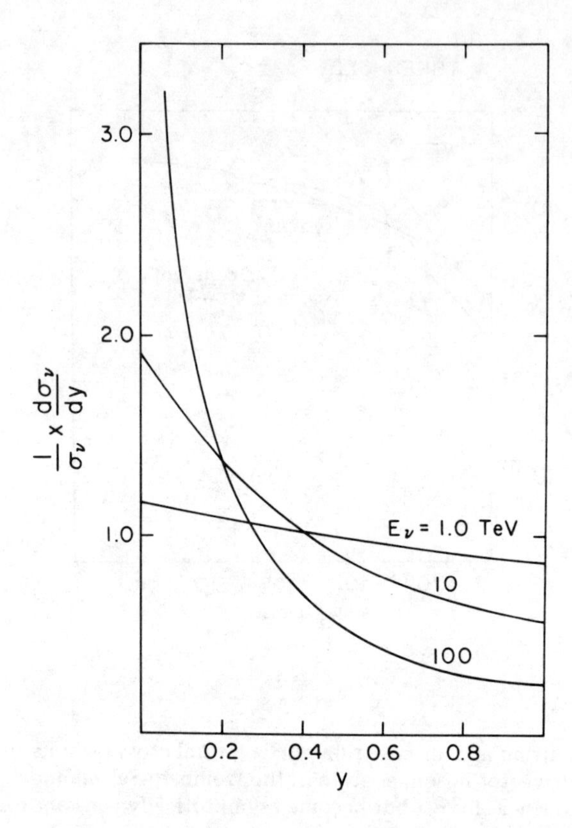

Fig. 6. The y distribution for neutrino-nucleon interactions at several energies shows increasingly elastic behavior. Note that growth with energy occurs at small y while the cross section at large y tends to a constant (Ref. 19).

these multi-muon states up to ~5 TeV at the 10^{-3} level and 20 TeV at the 10% level from the atmospheric flux alone. An interesting prospect for multimuon events (of small y) involves the use of events from a much larger volume, outside the array, but whose muons intersect the detector. Initial estimates are that the sensitivity thus extends upwards by about an order of magnitude.

Another interesting possibility is observation of the resonant interaction of electron antineutrinos with electrons (Glashow resonance) producing the W^- (Fig. 4d).[16-17] If there are significant extraterrestrial neutrino sources, these will be rich in electron neutrinos, and such interactions would be nicely detected for hadronic decays of the W.

Neutrino oscillations are another subject to which DUMAND can contribute. We shall not discuss this further here, but it seems likely that questions about oscillations over terrestrial distances will probably be settled prior to DUMAND operation.

The minimum estimated rates are indicated in Fig. 7. Extraterrestrial sources could add significantly beyond ~10^{14} eV. Clearly, there are a number of interesting observations that such a DUMAND could make. Let us go on and discuss some aspects of the detection techniques.

DISCUSSION OF DETECTION TECHNIQUES

A. Location of Detector

The favored locations for the detector are currently (though not finally) one of two deep basins near the Hawaiian Islands. The reasons for these selections have mostly to do with the availability of 5-km-deep water close to shore (~50 km), stable subsidence basins with good ocean environmental properties, and accessible high-quality support facilities on shore. The sites shown in Fig. 8 are northeast of Maui (upwind and therefore having rough surface conditions) and west of Hawaii, off Keahole Point in slightly shallower water, but in the lea of the big island. Initial measurements on currents, bottom conditions, and water clarity, particularly, are encouraging[5b] (see Table I). As indicated in Fig. 9, the array will be situated on the bottom, with an (as yet unspecified) mixture of optical and acoustic detectors deployed over a 1-km region, connected together and to shore by undersea cable. Power and commands emanate from shore and data returns over the same single (standard) coaxial cable. The dominant cause for choice of a great depth is indicated in Figure 10, which indicates that, while rates of through-going cosmic-ray muons would be > 10^4/sec at a 1-km depth (and the electronics always busy), at 5 km this rate would be down to about 10/sec, with only a tiny fraction of these muons (10^{-3}) producing significant cascades in the detector (and these are useful in themselves). The flux is confined to near-vertical directions.

Great depths (>2 dm) are also advantageous because of the <u>reduced</u> backgrounds of all other kinds, with the exception of the remnant light from radioactive decays in seawater and the thermal noise in the ocean (the levels of backgrounds above these limits are, of course, a matter of intense study,[5b] as indicated in Table I).

B. Detectors

Examples of detector modules are shown in Figs. 11 and 12. The favored optical detector utilizes wavelength-shifting fluor in a tube of index of refraction greater than water to trap Čerenkov radiation and guide it to phototubes. The design of acoustical modules involves more complex choices, which await both engineering trade-offs and Monte Carlo studies of arrays for further definition. It is important to observe that, in contrast to optical detectors, acoustic arrays may use phase information (because the

Table 1. Summary of preliminary DUMAND site studies underway or completed. Much more complete studies are planned (see Ref. 5b).

TEST	SCIENTISTS	INSTITUTIONS	MEASUREMENTS	STATUS/RESULTS
RUWS/ DUMAND	Learned, Wilkins, Blood, Lowenstine, Case, Gordon, Katayama, Peterson, Neefus, McLeod, Davisson, Peer	NOSC, UCI, UCSB, UH, U WASH, New England Aquarium	spectral absorption and scattering of light, bioluminescence; search, measure, and sample	instrumentation complete 1/78, deep dive 4/78
Bottom characteristics	Andrews	UH (OCEANOG)	bathymetry, cores, bottom photographic survey	some data 2 sites 5/77, bottom flat, good mooring
Physical characteristics	Harvey	UH (HIG)	currents; salinity, temperature profiles	some data 2 sites 10/77, homogeneity and currents good
Optical transmission	Zaneveld	OSU (OCEANOG)	attenuation profile at 650 nm	some data 2 sites 10/77; water very clear, virtually no nepheloid layer
Bottom seismicity	Odegard	UH (HIG)	ocean bottom recording seismometer	data 11/77 indicates Maui site activity
Acoustic noise profile	Blackinton	UH (HIG)	acoustic pulse record	in design
Barking Sands	Parvulescu, Blackinton, Bradner, Stenger, Wilkins, Peterson, Learned	UH (HIG), NOSC, SIO, UCI	characteristics of ocean impulsive noise	data 12/77, few natural bipolar pulses, but questions
BNL	Bowen et al. (+13 others)	BNL, HARV, U ARIZ, U WIS, UCI, SIO, UH, SYR, LSU, HIG	sound from groups of protons traversing water	complete; results consistent with thermoacoustic model
HARV	Sulak et al.	HARV	sound from 150 MeV stopping protons in various liquids, conditions	complete, as above, except anomaly in radiation at 6°C

Table I (continued)

TEST	SCIENTISTS	INSTITUTIONS	MEASUREMENTS	STATUS/RESULTS
Fermilab	Roberts et al.	FINAL, NRL, U ARIZ, UH, HARV, U WIS, UCI, SIO, SYR, NOSC, LSU, HIG	sound from 400 GeV proton cascades in water	scheduled for beam 8/78
LBL	Jones, Learned, Bradner et al.	LSU, UCI, SIO, U WIS, others	sound from heavy ions, various media, conditions	scheduled for ion beam 4/78
LSU	Jones et al	LSU	transducer studies, use of salt domes	in progress
UCI	Learned et al	UCI	transducer lab studies	in progress
	Bowen	U ARIZ	detailed pulse, potential	in progress
	Bradner	SIO	attenuation, hydrophones	continuing, 76 workshop
	Learned	UCI	attenuation, Monte Carlo	continuing
	Westervelt	BROWN	Fourier transforms, radiation	calculations in progress
	Hanish	NRL	summary of theory	notes 4/76
	Whitehouse	NOSC	arrays, imaging	76 workshop and in progress

Abbreviations for Table I.

SYR	Syracuse University	BNL	Brookhaven National Laboratory
U ARIZ	University of Arizona	BROWN	Brown University
UCI	University of California — Irvine	FNAL	Fermi National Accelerator Laboratory
UCSB	University of California — Santa Barbara	HARV	Harvard University
UCSD	University of California — San Diego (La Jolla)	HIG	Hawaii Institute of Geophysics
UH	University of Hawaii	LBL	Lawrence Berkeley Laboratory
U WASH	University of Washington (Seattle)	LSU	Louisiana State University
U WIS	University of Wisconsin (Madison)	NOSC	Naval Ocean Systems Center
OCEANOG	Department of Oceanography	NRL	Naval Research Laboratory
RUWS	Remote Underwater Work System	OSU	Oregon State University

SIO Scripps Institution of Oceanography, UCSD

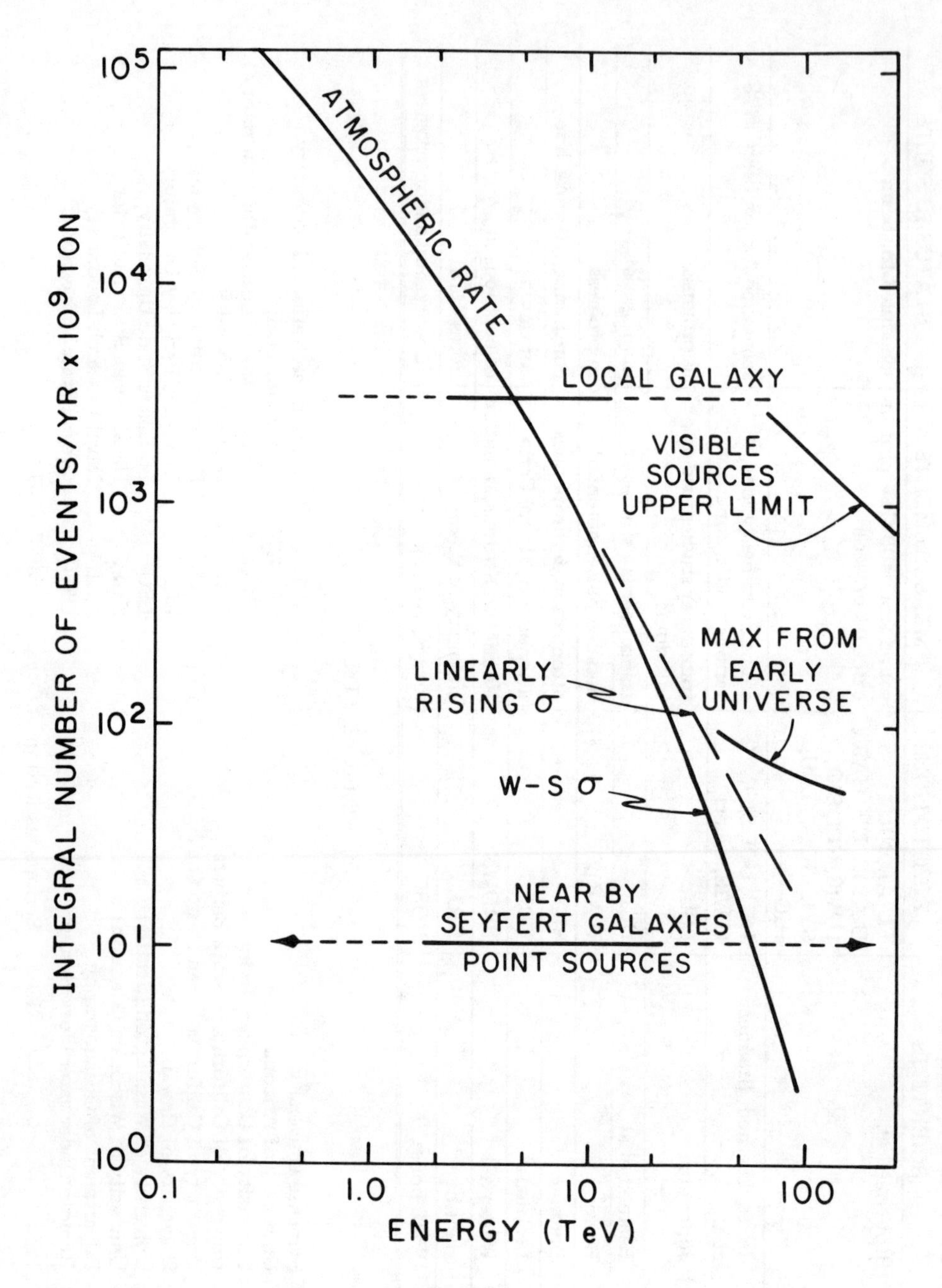

Fig. 7. Minimum rates in a 10^9 ton DUMAND array using atmospheric flux extrapolations and Weinberg-Salam cross sections. Several other source predictions are shown.

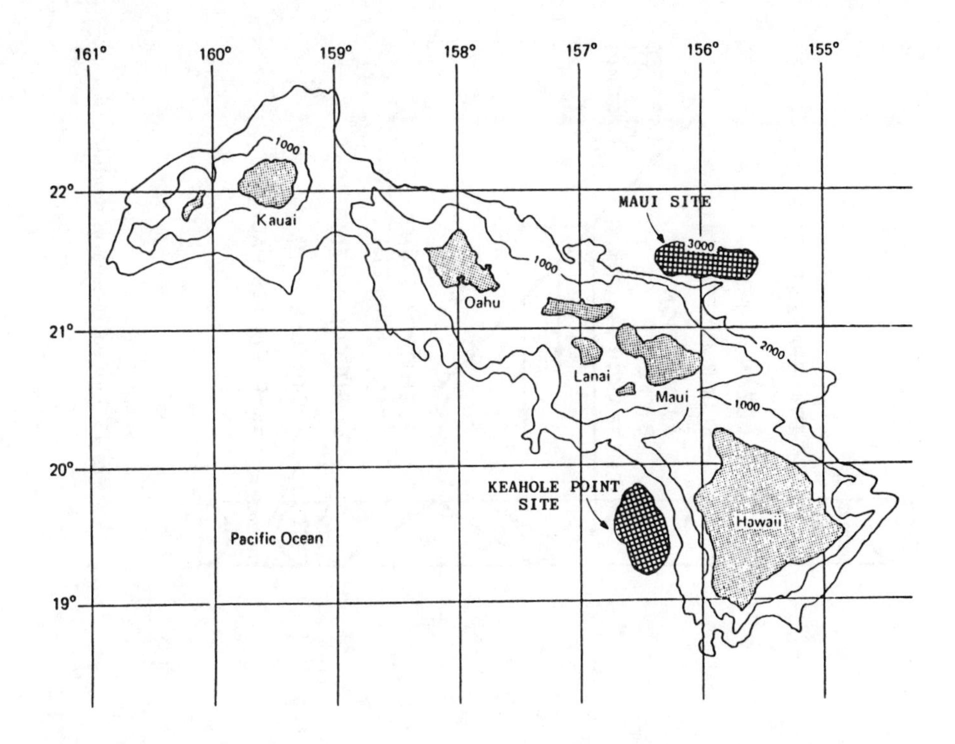

Fig. 8. Hawaiian Islands with two candidate DUMAND sites. Contours are in fathoms (1 fathom = 6 feet ≈ 1.8 m).

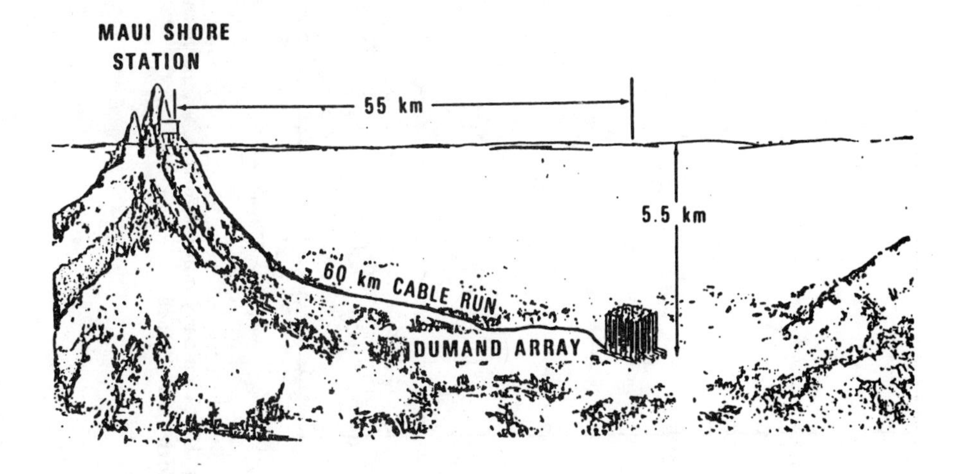

Fig. 9. Cross section with exaggerated vertical scale showing the array relative to the Maui Basin site.

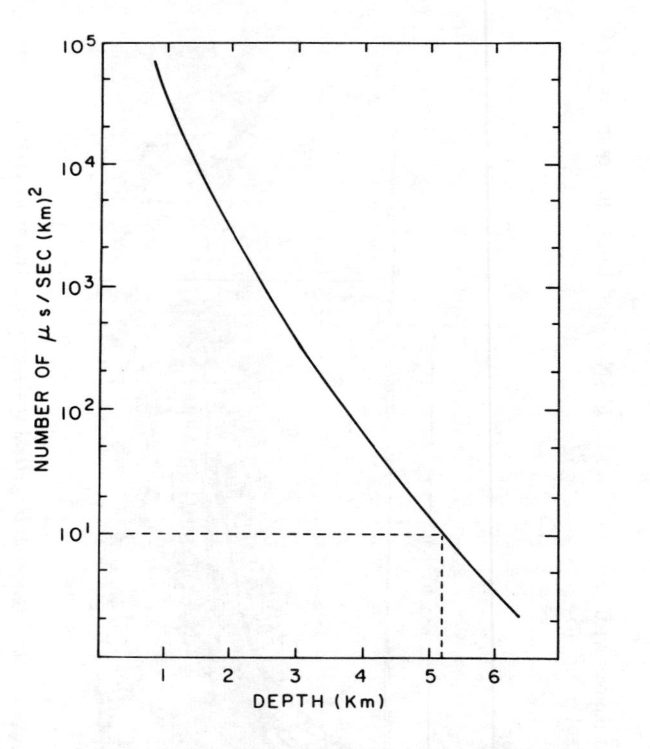

Fig. 10. Cosmic-ray muon depth vs. intensity through a horizontal surface.

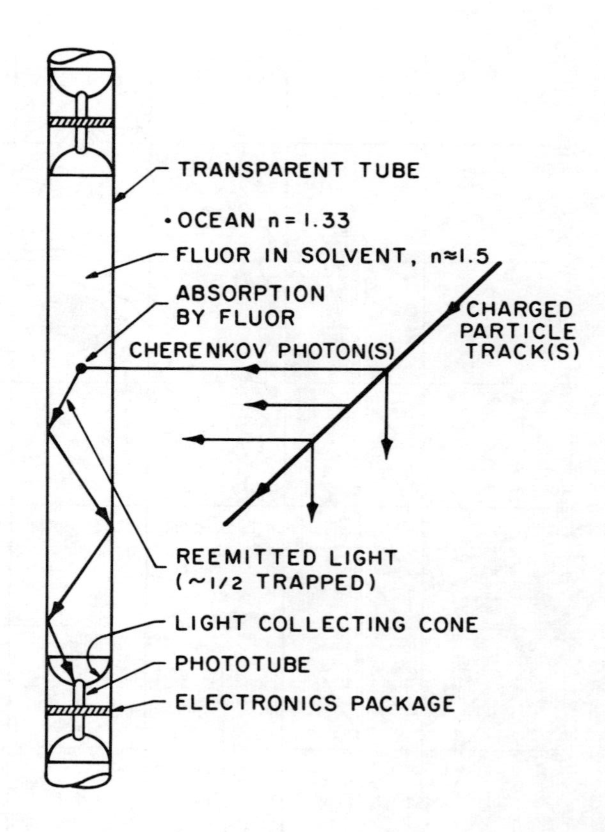

Fig. 11. Optical detection module.

source is coherent) and, thus, one array of N elements can simultaneously form N different beams, each of which has signal-to-noise ratio gain of N. This motivates one to "clump" acoustic detector elements into a few large modules.[3b] This has the advantage of making preassembly and deployment of the array relatively simple. When the number of elements reaches $\sim 10^4$, such a module can produce television quality images. Recent advances in theory and beam forming electronics permit scanning arrays with the number of components proportional to the number of elements[3b] (and not factorial or squared as one might initially suppose). It is not inconceivable to think of costs of the order of $10/element for these arrays.

We shall not discuss detectors here any further. For more details, the reader is referred to other sources.[1-5] It is important to note that no new technology is required for either the optical or acoustic approach, though for both of these the scale and location present a formidable ocean engineering challenge (if the experiment is to be conducted for a minimum expense).

C. Signal Characteristics

I would like to utilize the rest of this report to discuss various characteristics of the optical and acoustic signals. Let us first look at optical signals.

In the case of Čerenkov radiation, plotted as a function of wavelength in Fig. 13, we see that at successively greater distances the light becomes more peaked near 460 nm, in a band about 100 nm wide. It is this characteristic blue-green passband of water to which the wavelength shifting detector must be matched for maximum sensitivity at large distances. The curves are plotted for a 20-m attenuation distance (at 460 nm), which is typical of very clear ocean water, though observations have found values of up to 40 m (this is, incidentally, much clearer water than one can easily obtain, say, from lakes or wells). The dominant background will come from the light produced by radioactive decays of isotopes (particularly K^{40}) in the ocean water, which are constant throughout the ocean and unavoidable. Other sources, such as biological light, are at least not omnipresent, though one of the site selection criteria is, obviously, that the region chosen must be devoid of significant bioluminescence. (The time scale of biological flashes is measured in milliseconds and, thus, they will not produce false counts, though dead time or even complete temporary paralysis of the array could result from passage of a school of creatures emitting such light.)

Some signal-to-noise gain (~ 3) is realized by matching the detector absorption spectrum to the light at large distances (Fig. 13). One is not concerned with false coincidences produced by individual radioactive decays but by largely uncorrelated background which is equivalent to phototube noise. It is this background light that sets the detection threshold. Clearly, the value of the threshold energy for detection of a muon or cascade at some distance is a matter of economics. Muons of minimum ionizing energy loss rate are reliably detectable by strings of detectors (as in Fig. 11) at about 20 m, while cascades and muons of extreme energies ($\gg 1$ TeV) may be detected up to about 100 m, where the combination of attenuation and far field divergence produce a distance (r) dependence of the form

$$\frac{1}{r^2} e^{-r/\lambda}, \qquad (2)$$

where λ is the attenuation length of light in the band around 460 nm.

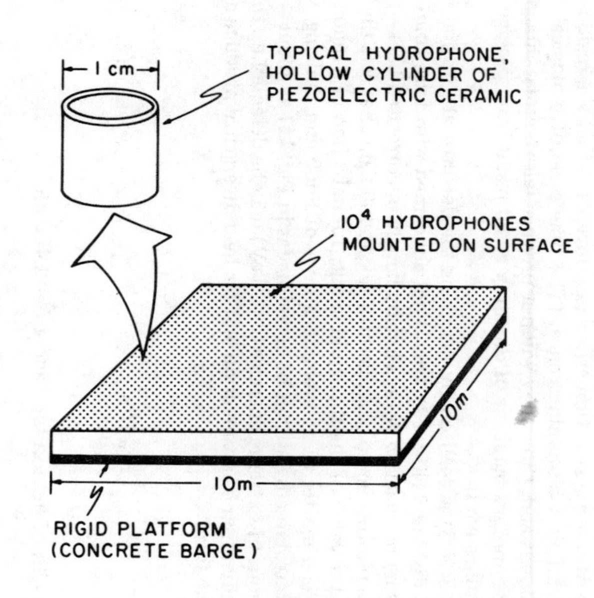

Fig. 12. Acoustic detection module.

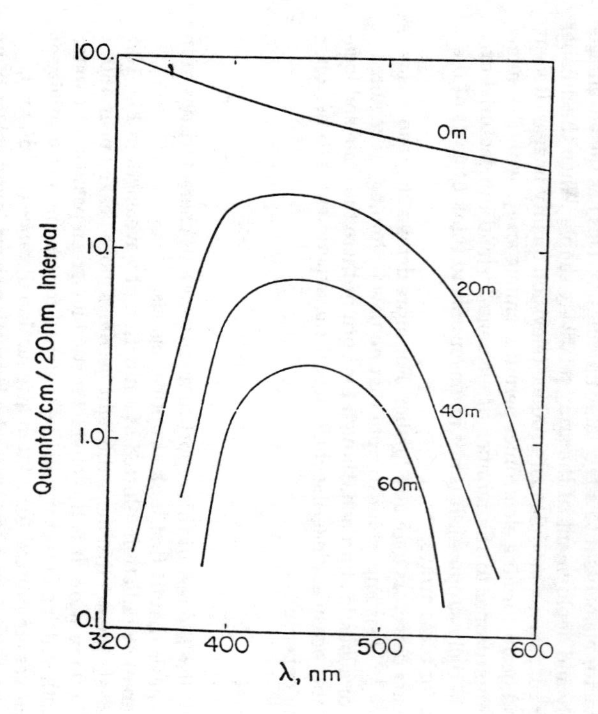

Fig. 13. Spectral distribution of Čerenkov light at various distances from a relativistic-charged-particle track in the ocean. Water with 20 m attenuation length at 460 nm is illustrated (ocean observations yield up to 40 m at this wavelength[2]).

For cascades, the shape of the contours of constant quanta per unit area define a volume inside of which the cascade may be detected, given some detector threshold sensitivity. Examples of this are shown in Fig. 14 for two energies and a photon flux contour of 200 quanta/m^2. One sees from this that a 10 TeV cascade is visible at this level at ~100 m in the direction of the Čerenkov cone and that the volume (the "detectable volume") within the contour surface is of the order of a million tons. The energy dependence of this detectable volume (V_D) is shown in Fig. 15a. The energy dependence (for a single detector) depends upon the 3/2 power of the energy, for energies roughly less than 100 GeV, due to the cubic increase in volume with distance in opposition to the $1/r^2$ intensity dependence. Exponential absorption of the light slows growth beyond this energy. Note that the situation is more complicated for an array of detectors (one must use a Monte Carlo program to evaluate possible geometries) and that the volume used in an analysis of an event may be much larger than that used for "detection" (triggering the recording of data).

The spectrum of the acoustic radiation from an infinite line[20] is indicated in Fig. 16 and is about proportional to the square root of the frequency up to the cutoff frequency. This cutoff is determined ultimately by the medium, but may be lower due to the finite dimensions of the heated region, as in the case of a cascade. As one goes out in distance, the lower part of the frequency spectrum steepens, becoming more nearly proportional to frequency at large distances (few km), while the high-frequency cutoff moves downwards (to ~100 kHz at 1 km).

The limiting noise background in the acoustic case is set by the thermal agitation of the ocean, at least above 10kHz, where the noise power per unit frequency interval rises as the square of the frequency. The incoherent noise in the deep ocean will be somewhat above this value (near the surface, it is dominated by wave noise), and there will be noise due to cetaceans and possibly other creatures. Certainly, the noise lessens as one goes to greater depth, but measurements of the actual levels (and their fluctuations) are yet to be obtained. It should be noted that in this case as well, natural sources (excluding the ones we seek) cannot give a false signal, but certainly can cause increased thresholds and array "busy" time.

Again, as in the optical case, temporal and spectral matching to the signal will produce gain in signal-to-noise ratios (factor of 4) plus rejection of biologically produced pulse trains. As opposed to the optical case, one can make further gains with "optics" (mentioned in III-B), which can further reject background by (at least in principle) not only angular but depth-of-field focusing. Certainly, the greatest sophistication in signal processing is going to be required because the signal is tiny by any measure. Without going into detail, it appears that the acoustic threshold for cascades is about 10^{14} eV (far field equivalent at 1 m in one hydrophone with 0 dB inherent maximum signal-to-noise ratio). For a practical detection module of 40 dB gain, this translates to a detectable volume of $>10^9$ m^3 at 10^{16} eV.[20]

The energy dependence, as indicated in Fig. 15b, is remarkably different for acoustical detectors, rising as the fourth power of the energy at first and then as the cube of the energy. The transition from near field (pressure proportional to $1/\sqrt{r}$) to far field (pressure proportional to 1/r) occurs at about 400 m in a direction transverse to the cascade axis (see Fig. 17). One can picture the shape of the detectable volume as a disk of thickness 10 m and diameter 800 m at that point. This directivity of the source (being a long thin coherent radiator) means that the observation of the acoustic disk gives the direction of the cascade to good accuracy (for astronomy). Sense of motion of the cascade can also be determined.[20] The (gentle) transition from far field to attenuation zone occurs at a few km (exact distance dependent upon energy, the cascade model, and acoustic attentuation in the deep ocean). Because of the bipolar nature of the

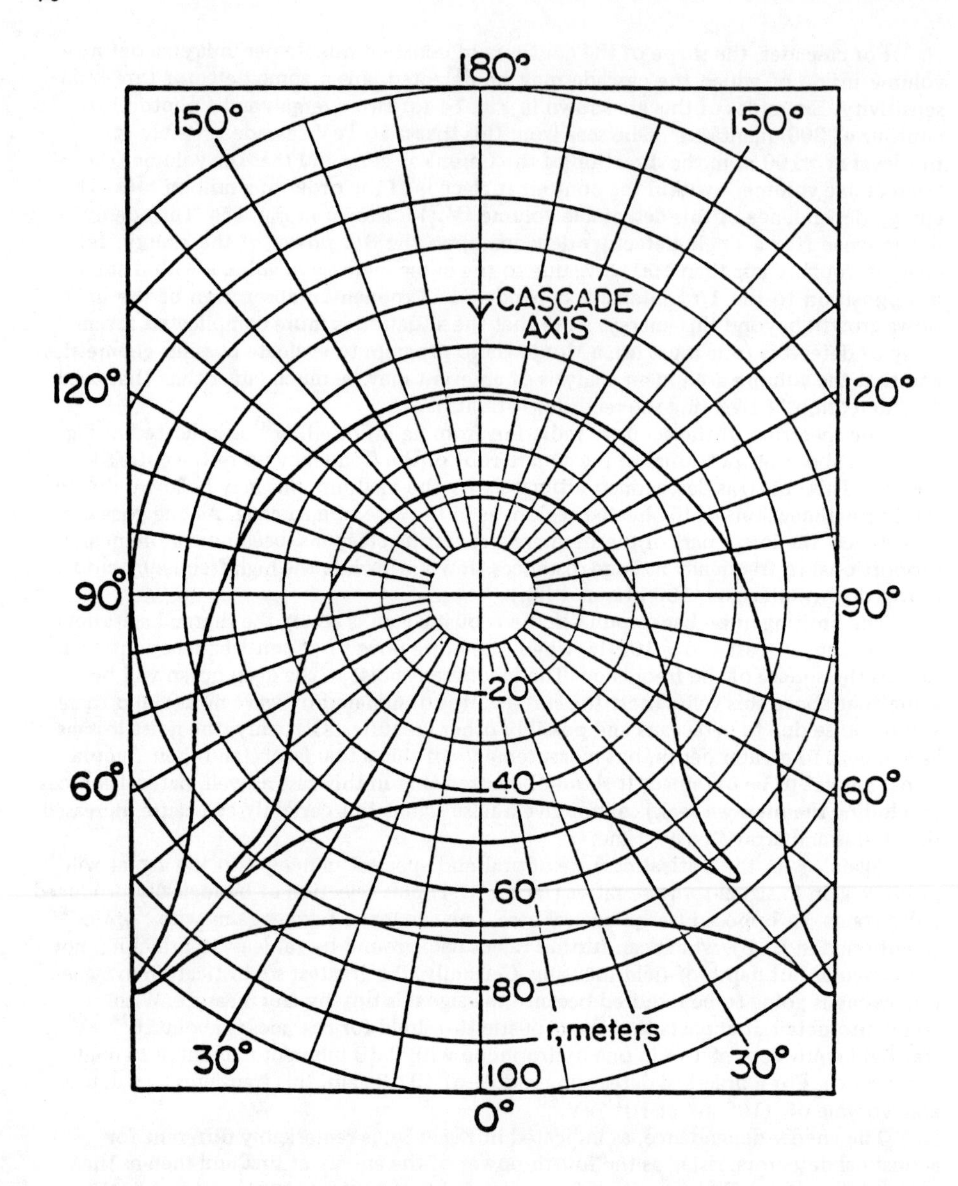

Fig. 14. Constant light-intensity contours for two energies. These also closely resemble constant signal-to-noise countours and the shape of the "detectable volume." The contours are for 200 quanta/m^2 and an optical attenuation length of 20 m at 460 nm.

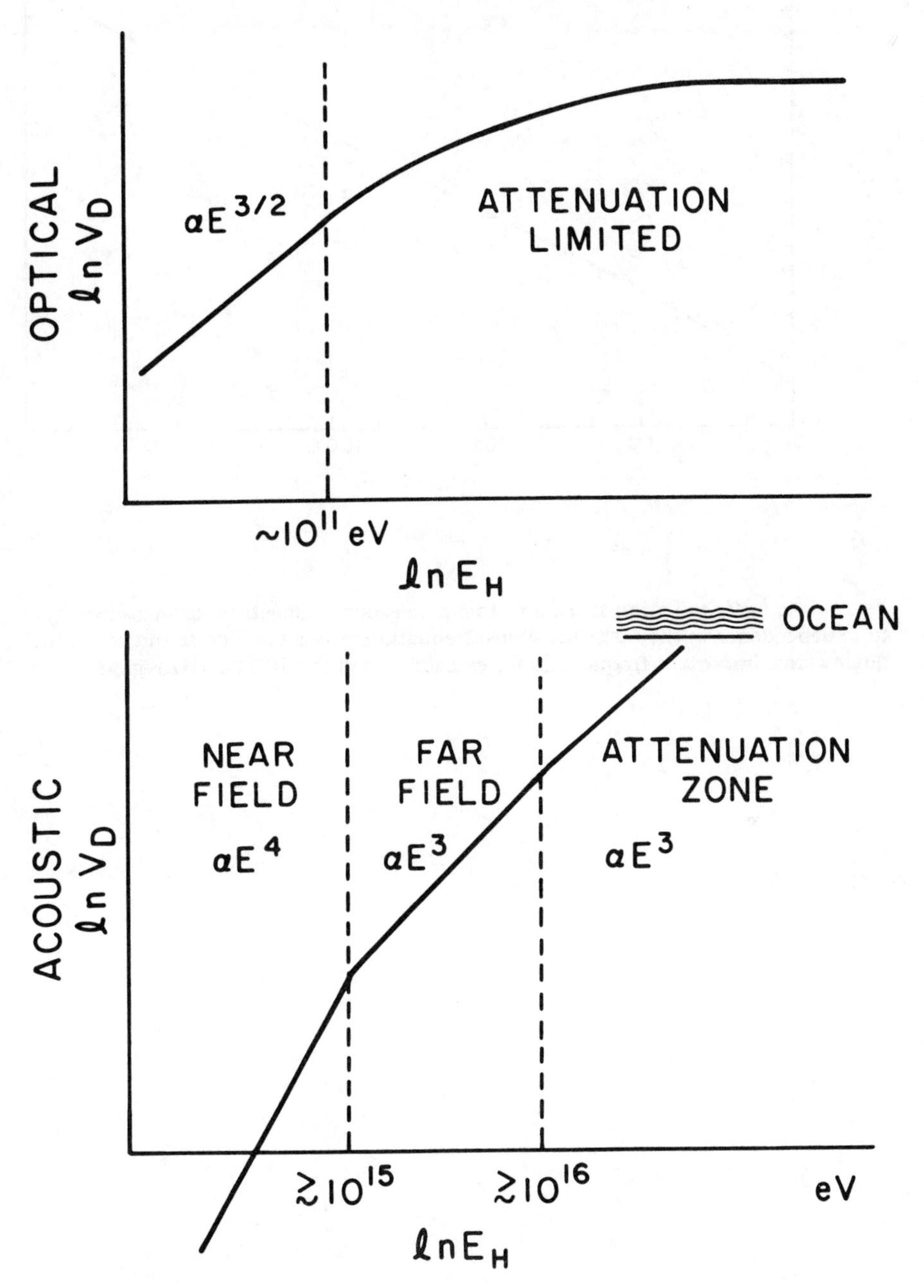

Fig. 15. Comparison of the dependence of detectable volume upon cascade energy for optical and acoustic detectors.

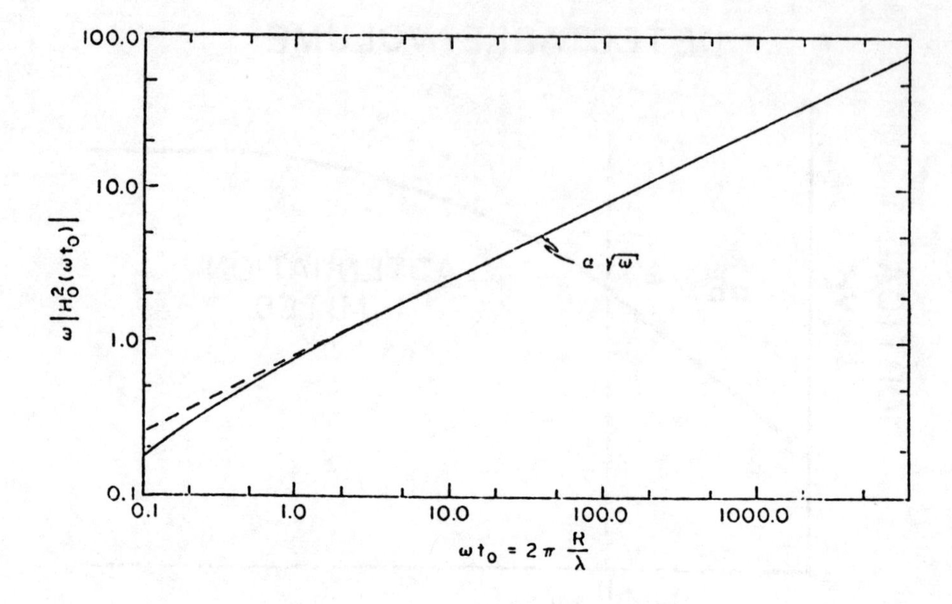

Fig. 16. Acoustic radiation spectrum from a line source. Spectrum changes from $\sqrt{\omega}$ to ω dependence in the far field, while attenuation moves the high frequency cutoff downwards. Important frequencies for cascades are in the 10-100-kHz region.

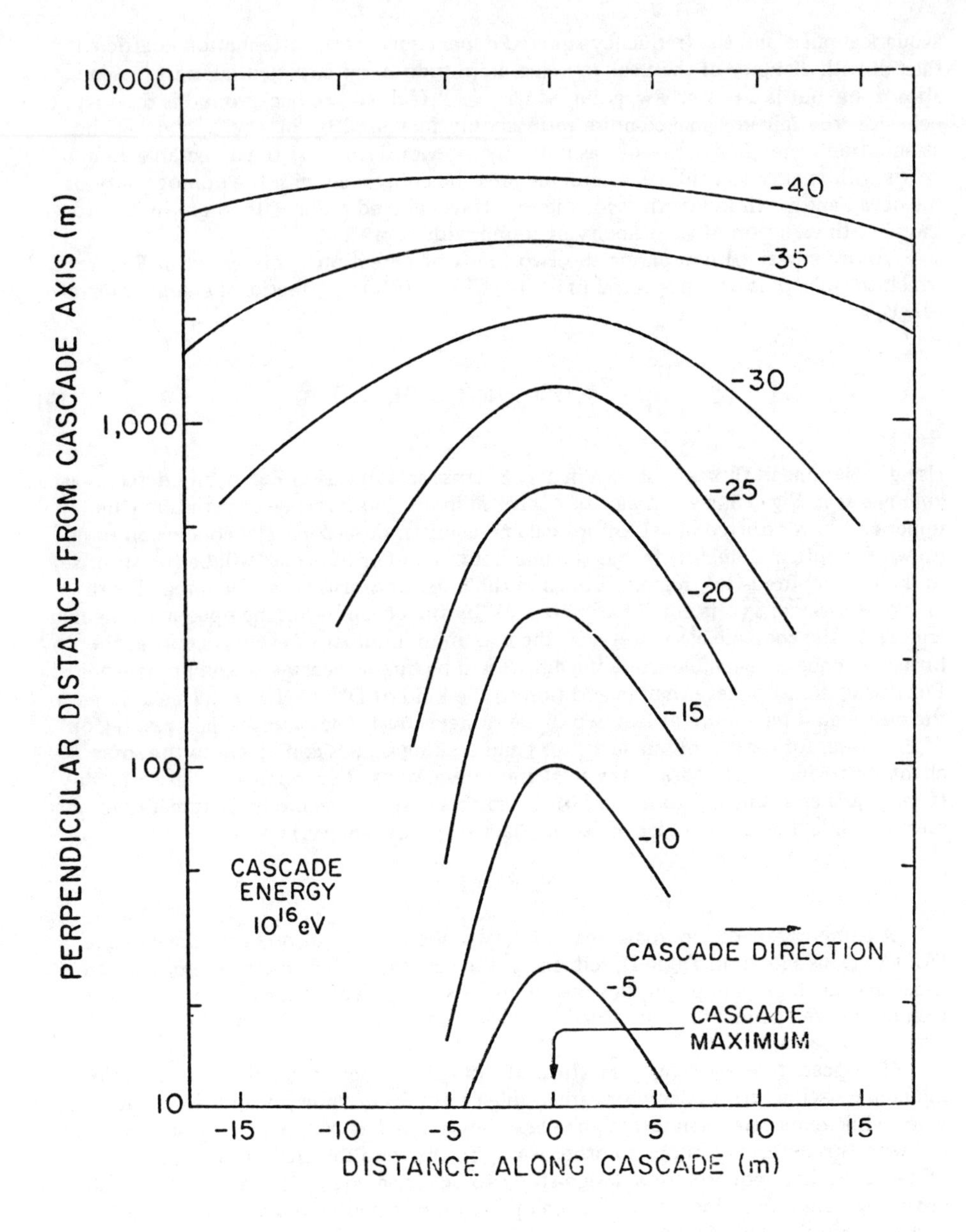

Fig. 17. Acoustic radiation from a nucleon cascade (isobars).

acoustical pulse and the frequency-squared dependence of the attenuation coefficient, the fall with distance of the peak pressure amplitude is not exponential in the attenuation zone, but is a power law, going as $1/r^{3/2}$.[20] If the noise background is thermal in behavior, the fall in signal-to-noise ratio is only increased by one-half power in the attenuation zone! The net result, as indicated previously, is that the detectable volume grows with energy to volumes where one must be concerned with the homogeneity of the ocean and its finite depth (acoustic rays travel curved paths with km radii, due to the smooth variation of the velocity of sound with depth).

Another way of comparing the two kinds of detection is presented in Fig. 18, which shows a plot of the spectral index (α) of rate (dR) per fractional energy interval (dE/E):

$$\frac{dR}{dE/E} = \Phi_\nu(E)\sigma_{\nu p}(E)V_D(E)E \equiv AE^{-\alpha}. \tag{3}$$

Using variations in fluxes as shown in Fig. 3, cross sections as in Fig. 5, and detectable volumes as in Fig. 15, we can plot this spectral index and its range of variation (the upper and lower dotted lines) for optical and acoustic detectors. The conclusion to be drawn for optical detectors is that the maximum number of events will be (no surprise) in the few-hundred-GeV region, though in the most optimistic case the integral with energy is divergent (gets cut off at $\sim 10^{17}$ eV by the ocean becoming opaque to the neutrinos). In the acoustic case, however, the maximum number of events occurs at the highest energies observable. Here the cutoff will be due to ocean size and homogeneity. For this reason, I believe that an addition to the kind of DUMAND we have explored in the past would be a more or less two-dimensional acoustic detector using a vast region of the ocean for target volume ($10^{15} m^3$) and seeking signals conducted in the sofar channel as evidence for cosmic-ray neutrino interactions at extreme energies ($\sim 10^{20}$ eV). (Going farther yet in size to the use of the earth's core, or the moon, as target does not seem possible due to geological noise and lack of homogeneity.)

SUMMARY

A summary of the comparisons of the two possible detection techniques usable in DUMAND, as shown in Table II, reduces to the statement that optical techniques are necessary for high-energy physics measurements in the TeV regime, while acoustic techniques are optimal for astrophysical observations at energies in the QeV (10^{15} eV) region and beyond.

It appears that by using a mixture of optical and acoustic detection, a cubic kilometer-sized neutrino detector is plausible (within economic possibilities). There are worthwhile experiments in ultrahigh energy physics and astronomy that can be carried out with this detector (and in no other way). To that end the DUMAND group has initiated funding requests for a design study to be carried out over a two-year period. Prospects continue to look good for the operation of a neutrino telescope in the deep ocean within the next decade.

Table II. Comparison of detector techniques.

OPTICAL: Best for lower energies and detailed measurements.

- Well-known lab technology.
- Fast coincidence possible.
- Source easily detectable < 100 m, very fast.
- Afterglow distinguishes EM from hadronic cascade.
- Detect muons.
- Small absolute "noise" level from K^{40} decays.
- Volume limited by water attenuation.
- Signal $\propto \frac{1}{r^2} e^{-r/\lambda}$.

ACOUSTIC: Best for superhigh energies and vast target volumes.

- New technique, but largely well-known technology.
- Electronics may be relatively slow.
- Inherently small signal with unique signature.
- Deployment in modules with many elements, easy.
- Background noise at depths above thermal, needs further study.
- Volume limited by inhomogeneities of ocean.
- Signal $\propto \frac{1}{\sqrt{r}} \rightarrow \frac{1}{r} \rightarrow \frac{1}{r^2}$.

  (Effect of attenuation is a power law.)

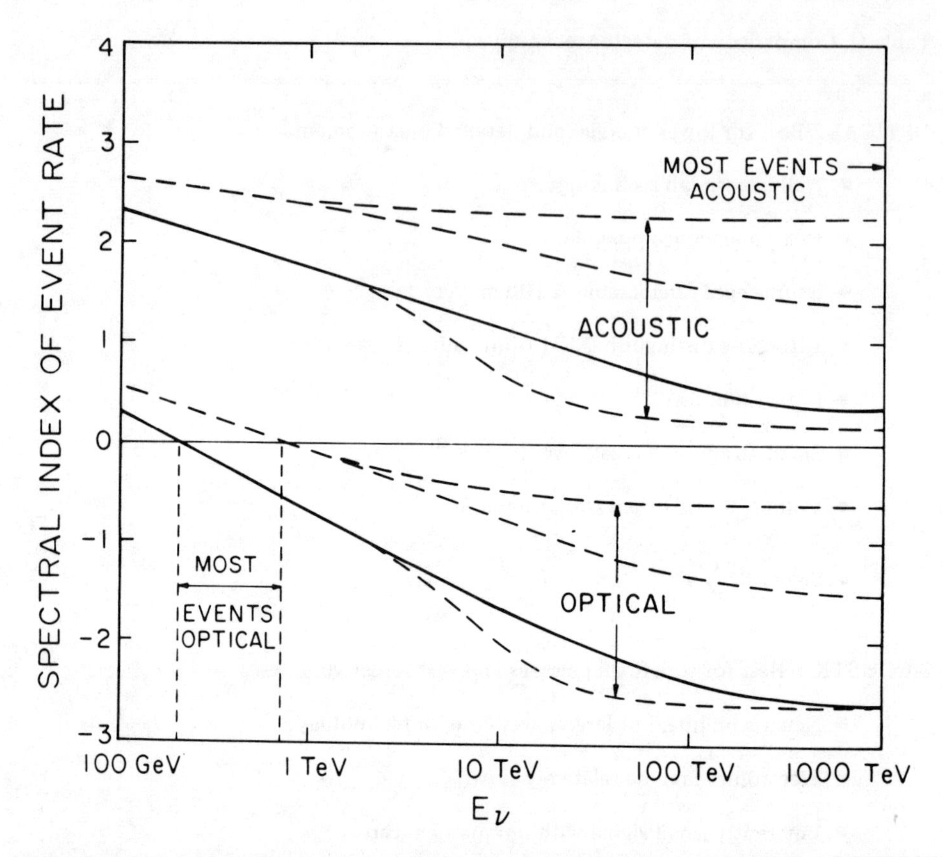

Fig. 18. A comparison of the spectral index of the event rate in a DUMAND array with optical and acoustic detectors. The dotted lines represent extreme assumptions. Both optical and acoustic curves will turn downwards at some (off-scale) energy, due to the finite depth and homogeneity of the ocean. Note maximum event rate for optical detectors will be in the 100-GeV to 1-TeV range while acoustic detectors have the remarkable property of having the highest rate at the highest observable energy.

FOOTNOTES AND REFERENCES

1. Proceedings of the 1975 DUMAND Summer Study, Bellingham, Washington, edited by P. Kotzer (Western Washington State College, Bellingham, Washington, 1976).
2. Proceedings of the 1976 DUMAND Summer Workshop, Honolulu, edited by A. Roberts (Fermilab, Batavia, Illinois, 1977).

3a. A. Roberts, Report on the 1977 DUMAND Summer Workshop, Moscow, 1977, Office of Naval Research (London Branch), ESN 31-9, 1977 (unpublished), p. 370.

b. Proceedings of La Jolla DUMAND Workshop on Data Processing, Scripps Institute of Oceanography, 1977, edited by H. Bradner (Scripps Institute of Oceanography, La Jolla, California, 1978).

c. Proceedings of the San Diego DUMAND Workshop on Ocean Engineering, 1977, edited by G. Wilkins (in preparation).

4. A. Roberts of Fermilab is the Director of the Workshop and will edit the proceedings. Dr. Roberts is also Secretary to the DUMAND Organization and should be contacted in order to receive DUMAND material, newsletters, etc.

5a. H. Blood, H. Bradner, J. Learned, F. Reines, A. Roberts, and G. Wilkins, DUMAND Design Study Proposal (University of California, San Diego, Scripps Institute of Oceanography, La Jolla, 1978), Vols. I-III.

b. J. Andrews, G. Blackinton, R. Harvey, E. Kampa, V. Peterson, and R. Zaneveld, Evaluation of Hawaiian Deep Water Sites for the DUMAND Project (University of Hawaii, Honolulu, 1978).

6. J. G. Learned, in Neutrinos-78, proceedings of the International Conference on Neutrino Physics and Neutrino Astrophysics, West Lafayette, Indiana, 1978, edited by E. C. Fowler (Purdue University, 1978), p. 895.
7. See also the pertinent contributions to Neutrino '77, proceedings of the International Conference on Neutrino Physics and Neutrino Astrophysics, Baksan Valley, U.S.S.R. 1977, edited by M. A. Markov (Nauka, Moscow, 1978), and to Neutrinos-78.

8. At the highest energies ($\gg 10^{12}$eV), the cascade energy distribution becomes increasingly dominated by electron neutrino interactions. Some cross section measurements can be made, and it may be possible to learn something from the cascade shape itself.
9. If we use the standard Bjorken scaling variables $x = Q^2/2m_p E_H$ and $y = E_H/(E_H + E_\mu)$, the resolution function for these variables can be shown to be

$$\left(\frac{\delta x}{x}\right)^2 = 4\,\frac{[\delta(\theta_\mu + \theta_H)]^2}{(\theta_\mu + \theta_H)^2} + y^2\left(\frac{\delta E_\mu}{E_\mu}\right)^2 + (1-y)^2\left(\frac{\delta E_H}{E_H}\right)^2,$$

$$\left(\frac{\delta y}{y}\right)^2 = (1-y)^2\left(\frac{\delta E_H}{E_H}\right)^2 + (1-y)^2\left(\frac{\delta E_\mu}{E_\mu}\right)^2.$$

In certain kinematic regions ($y \rightarrow 1$), the resolution function is less dependent upon E_H and E_μ resolution; then a relatively poor measurement of those quantities can still yield a good measurement of x or y. For a good x measurement, the muon angle must be well determined. This should be the case in a DUMAND detector, since the muon would be observed over several hundred meters of path length. (For further details, see D. Cline, these proceedings.)

10. Actually the rate in a 10^9-ton detector is comparable to that at Fermilab in an early 100-ton detector at similar energies. The difference is, of course, that the cosmic-ray detector has a spectrum of events going beyond the machine energy, usefully to $>10^{14}$ eV.
11. Notice that extraterrestrial sources would produce a significantly different observed y distribution if the source were antimatter (assuming, as we are, that the detector is able, at least roughly, to measure muon and cascade energies and angles).
12. S. Margolis and D. Schramm, Proceedings of the Fifteenth International Conference on Cosmic Rays, Plovdiv, 1977 (Bulgarian Academy of Sciences, Plovdiv, Bulgaria, 1977), Vol. 6, p. 244.
13. S. Margolis, D. Schramm, and R. Silberberg, University of Chicago, Enrico Fermi Institute, preprint No. 77-39, 1977.
14. D. Eichler, "Ultra-High Energy Neutrino Emission from Cosmic Ray Sources," University of Chicago, Enrico Fermi Institute preprint, October 1977 (unpublished).
15. R. Silberberg and M. M. Shapiro, Proceedings of the Fifteenth International Conference on Cosmic Rays, Plovdiv, Vol. 6, p. 236.
16. V. S. Berezinsky and G. T. Zatsepin, Usp. Fiz. Nauk 122, 3 (1977) [Sov. Phys. Usp. 20, 361 (1977)].
17. V. S. Berezinsky, in Proceedings of the Fifteenth International Conference on Cosmic Rays, Plovdiv, Vol. 6, p. 231.
18. See also V. S. Berezinsky, in Neutrinos-78, p. 23; D. N. Schramm, in Neutrinos-78, p. 87.
19. T. Gaisser and A. Halprin, in Proceedings of the Fifteenth International Conference on Cosmic Rays, Plovdiv, Vol. 6, p. 265.
20. J. G. Learned, "Acoustic Radiation of Charged Atomic Particles in Liquids, An Analysis," University of California, Irvine, preprint UCI Physics No. 77-14 (submitted for publication in June, 1978).

THE ACOUSTIC DETECTION OF HADRONIC SHOWERS INDUCED BY COSMIC NEUTRINOS*

L. R. Sulak
Harvard University, Cambridge, Massachusetts 02138

ABSTRACT

A detectable sonic signal is produced by charged particles while traversing a fluid medium. This phenomenon could be exploited in an inexpensive shower detector for the massive ($\gtrsim 10^4$ ton) neutrino detector necessary if one wishes to detect neutrinos at large distances from the next generation of high energy accelerators, e.g., the Fermilab energy-doubler. It could also be used in the shower calorimeter of a massive ($\gtrsim 10^9$ ton) deep underwater detector of astrophysically produced neutrinos. This paper[1] discusses experiments exploring the global characteristics of both the acoustic-generation mechanism and the radiation pattern. The results of the experiments are consistent with a simple thermal model for the transformation of the energy of moving charged particles into acoustic energy.

The production of sound by particles traversing solid matter has been demonstrated[2] in aluminum and piezoelectric material. However, the quantitative interpretation of the results is difficult due to the small size of the targets and the complicated acoustic behavior of solids. The simpler acoustic properties of liquids make them better suited to a systematic investigation of the sound-generation mechanism.

In liquids, sound could be generated by charged particles via several mechanisms. These include the generation of a simple acoustic signal by (1) the adiabatic expansion caused by instantaneous heat deposition,[3,4] (2) the implosion of microbubbles produced by the ionizing particles,[5] and (3) molecular dissociation.[6]

Let us consider the sonic signal produced by a beam idealized as a long, thin, uniformly heated cylinder of diameter d and length L (see Figure 1). Naïve wave considerations[3,4] suggest that the sonic wave produced by the instantaneous expansion of this hot rod should have the following properties: (1) The half-wavelength $\lambda/2$ is on the order of the diameter d of the energy deposition. (2) The fundamental frequency f of the sound emitted is, therefore, c/2d, where c is the speed of sound (1.5 mm/μsec in water) and the period $\tau = 1/f$. (3) For observation angles $\theta \leqslant \lambda/L$, the sonic wave from the cylindrical acoustic antenna is coherent in the near field, since $L \gg d$. (4) This coherency and a $1/\sqrt{R}$ falloff of the sonic pressure signal P, where R is the observation distance, extend to a near-field limit $A = L^2/\lambda$. For observation distances beyond A (the far field), the pressure signal decreases as 1/R.

Assuming a simple thermal mechanism, dimensional analysis suggests that the pressure signal should be proportional to the volume coefficient of expansion K divided by the heat capacity of the medium C_p. The signal should be linearly related to the total energy deposited, E, and, in the near field of an acoustical antenna, should be proportional to $1/\sqrt{LR}$,

$$P(\text{dynes/cm}^2) = (K/C_p)\,(E/\sqrt{LR})\,M.$$

Dimensional analysis does not determine the model-dependent term M (units of $\sec^{-2}$). Two models have been proposed. Dolgoshein and Askarian[4,5] assume uniform heating

*This work was supported in part by the U.S. Department of Energy under Contract No. EY-76-C-02-3064. *000 and by the National Science Foundation.

ISSN:0094-243X/79/520085-15$1.50

within the hot rod, which produces an f^2 dependence in M. This is multiplied by an angular-dependent term common in problems with cylindrical geometry,

$$M = (f^2/2)(\sin x)/x ,$$

where $x = (L/2\lambda)\sin\theta$. An independent calculation by Bowen[3] predicts a signal with characteristics similar to those described above.

The importance of the thermal mechanism in the production of the sonic signal can be directly tested by measuring the absolute value of the sound pressure and by verifying its dependence on E and K/C_p. The K/C_p dependence can be determined either by concentrating on the strong temperature dependence of K in a single material, or by surveying materials of different K/C_p. In the latter method, one must correct for the simultaneous change of density, speed of sound, and acoustic impedance.

Experimental investigations of the acoustic signal from protons traversing or stopping in liquids have been performed in three experimental situations. The first one utilized a large tank of water (with dimensions $\gg$ d or L), to avoid possible confusion of the primary signal with reflections from the tank walls. This experiment, performed on the 200 MeV linac at Brookhaven National Laboratory,[7] observed heavily ionizing protons stopping in water (range = 30 cm). Deposition times ranged from 3 to 200 μs and total energy depositions from 10^{19} to 10^{21} eV. The beam could not be tuned to lower total-energy depositions. The diameter of the beam was fixed at 4.5 cm.

Similar experiments have been done at the 158-MeV cyclotron at Harvard University. Here, the energy deposition could be decreased to 10^{15} eV, allowing one to approach the threshold levels of sound production. Protons of this energy have a range of 16 cm. However, the pulse duration was limited to a minimum of 50 μsec. This period is long compared to the transit time across the beam diameter and, thus, it dominated the time structure of the sonic signal. The cyclotron experiments used smaller vessels, but with dimensions $\gg \lambda$. The easy access to the experimental area at the cyclotron permitted the rapid target changes necessary to explore the dependence upon chemical composition, temperature, and pressure.

A third set of experiments was carried out in the 28-GeV proton fast-extracted beam (FEB) at BNL. As in the BNL linac experiment, the beam could not be tuned below energy depositions of 10^{19} eV. Typically, 3×10^{11} protons per pulse traversed 20 cm of water. The beam diameter was variable between 5 and 20 mm, and the beam deposition time was short (2 μsec). In contrast to the cyclotron experiments, the transit time across the diameter of the beam dominated the time structure of the sonic signal.

The transducers used to detect the acoustic signal are of three varieties. One type is a standard Navy hydrophone and amplifier configuration, with a gain of 0 to 110 db and a minimum sensitivity of −80 db re 1 volt/dyne/cm^2, uniform between 1 KHz and 150 KHz. A second type of hydrophone is sensitive up to 1 MHz, due to its small diameter. However, its small volume limits its sensitivity to −115 db re 1 volt/dyne/cm^2. A third type of hydrophone was manufactured with a preamp built into the hydrophone housing. The system has very low self-noise and recently was calibrated by the Navy.

Typical signals from the two hydrophones in the linac experiments are shown in Fig. 2 for a beam-spill time (10 μsec) short compared to the sound transit time across the beam diameter. The scope traces exhibit remarkably simple bipolar pulses, followed much later (with the appropriate time delay) by reflections from the bottom of the tank. The half-period of the bipolar signal (30 μsec) is that expected from the sound transit time across the beam diameter (4.5 cm). The time delay between the signals (150 μsec) corresponds to the difference (~ 20 cm) in path lengths between each of the microphones and the source. We have confirmed that the first half-period of the signal is compressional (the two hydrophones are wired 180° out of phase).

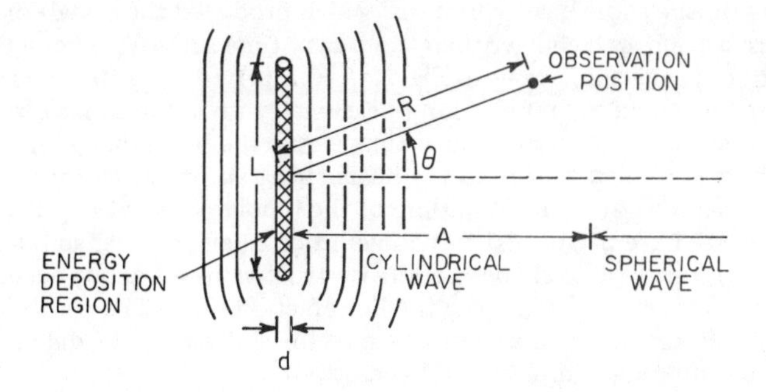

Fig. 1. Sonic signal produced by a charged-particle beam.

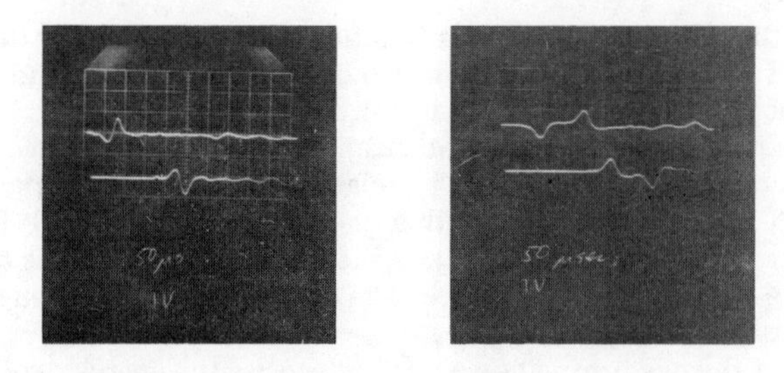

Fig. 2 (left). Typical hydrophone signal for short beam-spill time.

Fig. 3 (right). Typical hydrophone signal for long beam-spill time.

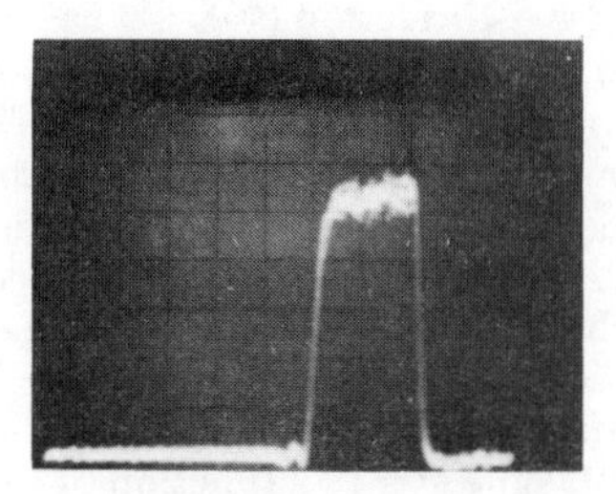

Fig. 4. Time distribution of beam spill producing the sonic pulse of Fig. 4.

In contrast to the short beam durations which produced the signals discussed above, a 100-μsec beam spill (longer than the transit time (30 μsec) characteristic of the diameter) produced the sonic pulses in Fig. 3. A null period of ~ 50 μsec separates an initial compression and a final rarefaction pulse, each with a duration substantially longer than that appropriate to the transit time across the beam diameter. Fig. 4 illustrates the uniform time distribution of the beam spill that produced this sonic pulse. The resultant signal is a convolution in time of the bipolar pulse of Fig. 2 with the beam profile of Fig. 4. We have performed this convolution in a computer simulation and reproduced the observed pulse shape. Totally destructive interference occurs after the initial compression wave and before the final rarefaction wave. The width of the compression wave is plotted in Fig. 5 as a function of the spill time. The data is fit well by the results of the computer simulation (also plotted).

Fig. 6 shows the peak signal amplitude as a function of spill time for fixed beam current. For beam durations less than $d/c = 30$ μsec, the pressure is linearly related to the spill time (and therefore to the total energy deposited). For spill times > 30 μsec, the peak signal reaches a maximum due to the destructive interference discussed above. The transition from the linear portion of the curve to saturation is fit well by the computer simulation.

Fig. 7 shows the signal amplitude as a function of the total energy deposition for spill times $\ll d/c$. It is linear over more than two orders of magnitude and independent of the spill time (at the same energy deposition).

In the FEB experiments, we varied the beam diameter and measured the change in period and amplitude of the sonic pulse. The relationship between the period and the diameter, shown in Fig. 8, is consistent with a linear dependence. Fig. 9 shows the relationship between the signal amplitude and the beam diameter d. The thermal model[4,5] predicts a $1/d^2$ ($\propto f^2$) dependence. The data fit a $1/d$ dependence better, but could accommodate a $1/d^2$ relationship.

Experiments at the cyclotron allowed us to decrease the energy deposition to a minimum of 10^{15} eV. Fig. 10 shows the signal amplitude as a function of energy deposition recorded in this low-energy regime. Within a factor of two, it is consistent with a linear extrapolation (of Fig. 7), at 5 orders of magnitude less than in the experiments with the linac. For this comparison, the same hydrophones were used and corrections for the signal dependence on distance, beam-spill time, beam radius, and deposition length were applied.

In Fig. 11, we display the volume expansivity and heat capacity of some representative fluids.[8] Fig. 12 shows the measured dependence of the acoustical amplitude on K/C_p for five of these materials. A correction ($\propto L^{-1/2}$) has been applied, since the proton range (and therefore deposition length) differs among the materials. The acoustic signal has been normalized to the total beam energy. The data in Figure 12 exhibit a linear dependence of the acoustic amplitude on K/C_p over more than an order of magnitude. These results support a dominant thermal mechanism for the generation of the sonic wave. In particular, there is no evidence for a signal enhancement in CCl_4 above that expected from a thermal mechanism. This had been predicted[6] if molecular dissociation or ion formation were a significant producer of sound.

The evidence for the thermal origin of the sonic pulse is strengthened by measurements of the signal amplitude from water as a function of temperature. Fig. 13 displays the rapid, nearly linear variation of K as a function of temperature.[9] Also displayed is the observed signal as a function of temperature, showing the expected variation. At 4°C water reaches its maximum density and for temperatures < 4°C K becomes negative. (An increase in temperature will cause contraction.) For water, C_p is constant to within 0.5% for the range of temperature that we explore. Thus, the temperature dependence of the sonic signal in water separates K from the K/C_p dependence verified above.

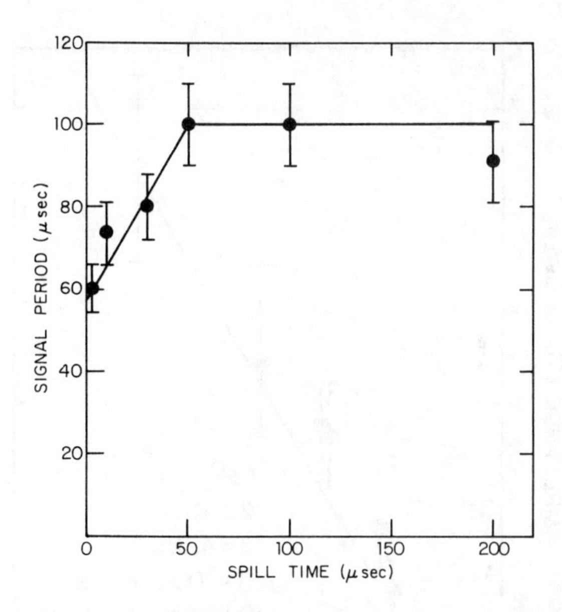

Fig. 5. Width of signal as a function of spill time.

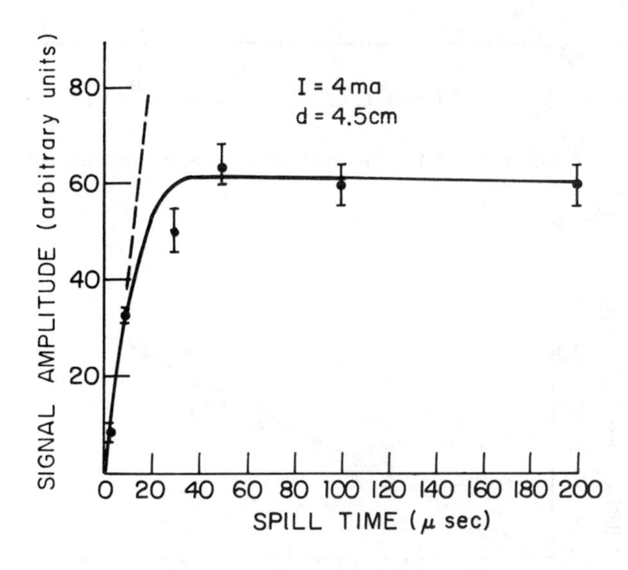

Fig. 6. Peak signal amplitude as a function of spill time.

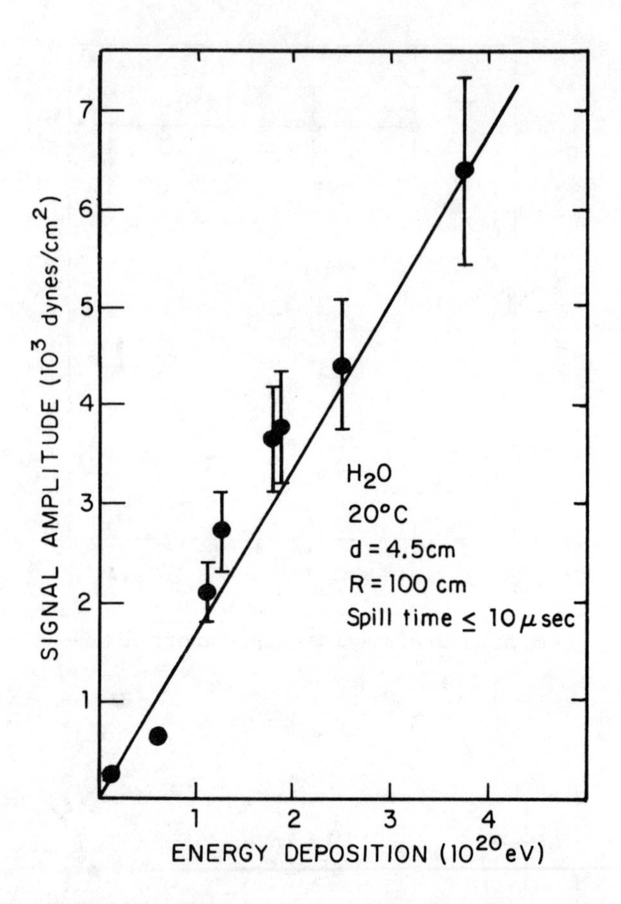

Fig. 7. Signal amplitude as function of total energy deposition.

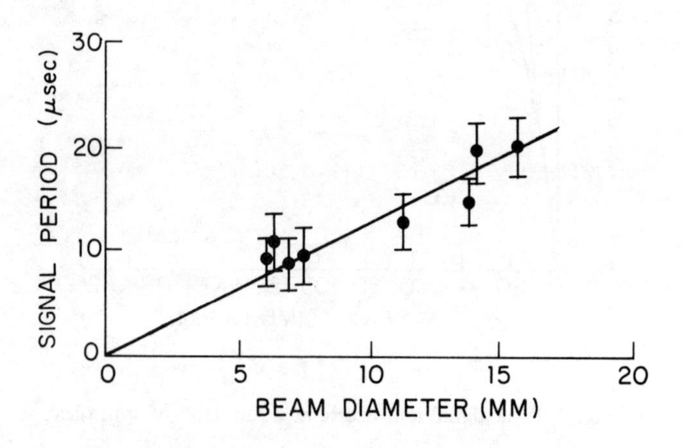

Fig. 8. Signal period as function of beam diameter.

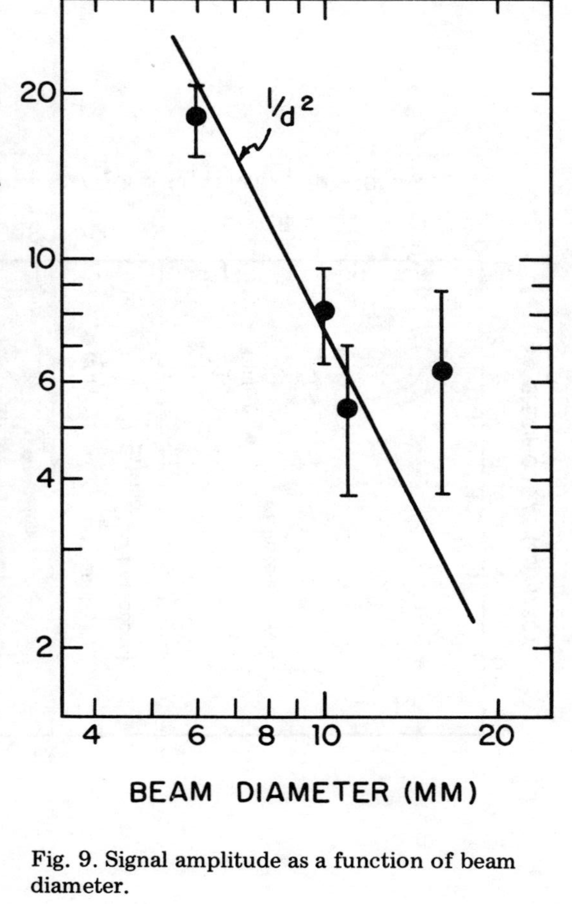

Fig. 9. Signal amplitude as a function of beam diameter.

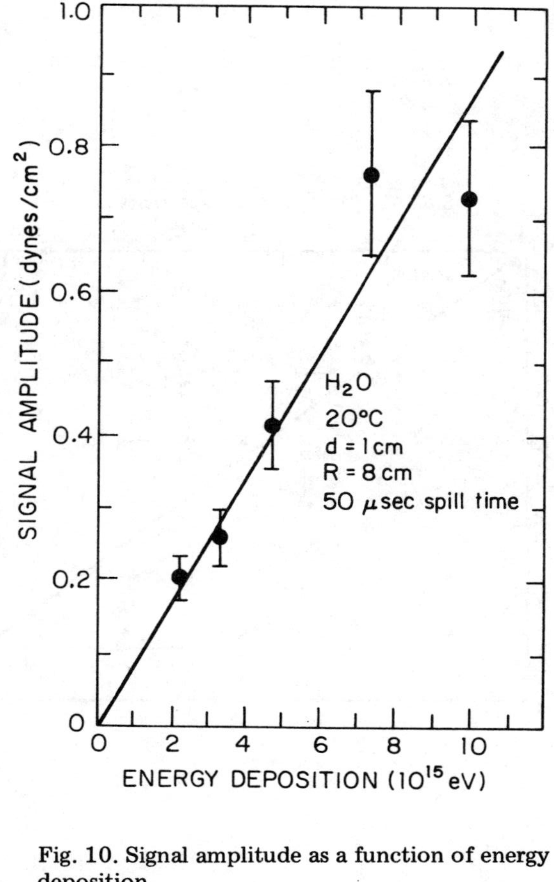

Fig. 10. Signal amplitude as a function of energy deposition.

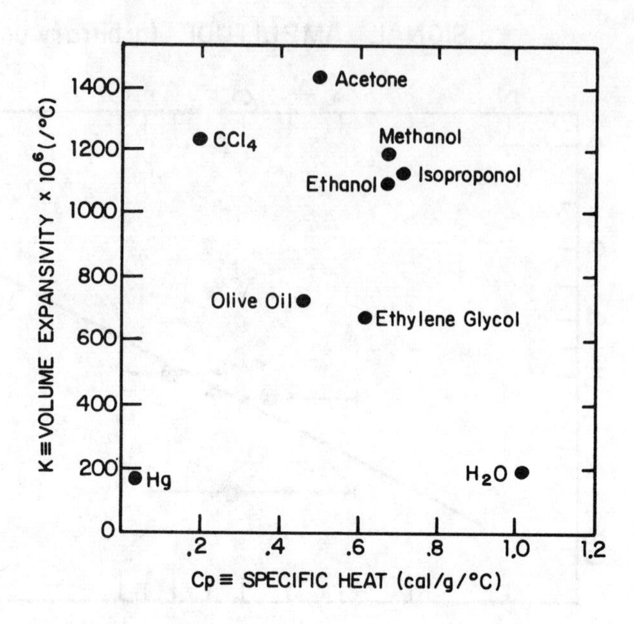

Fig. 11. Volume expansivity K and heat capacity C_p of some representative fluids.

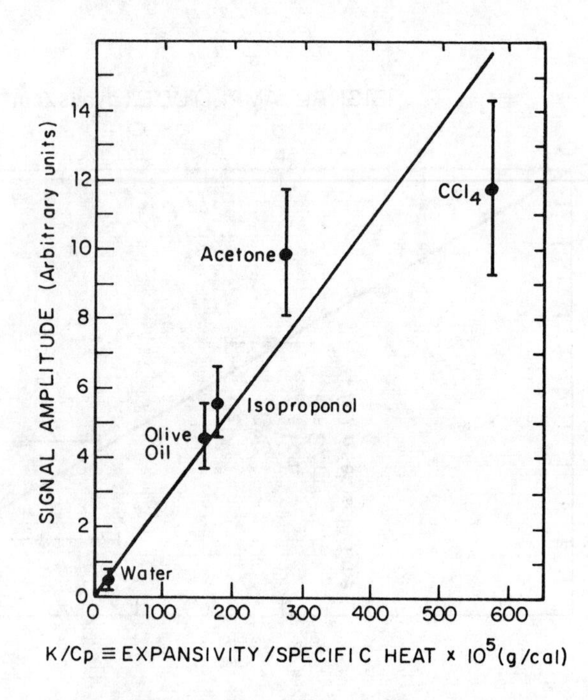

Fig. 12. Measured dependence of acoustical amplitude on K/C_p.

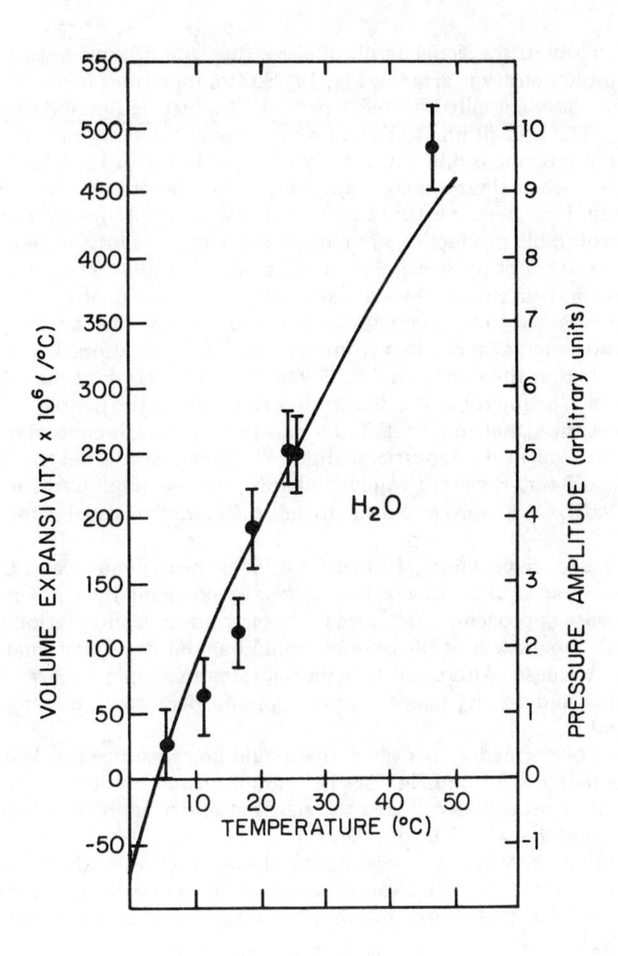

Fig. 13. Variation of volume expansivity K with temperature, and observed signal as a function of temperature.

A detailed survey of the signal amplitude as a function of temperature in the vicinity of maximum density is given in Fig. 14. For temperatures below 6°C, the observed signal in water actually becomes inverted: the first half-period of the signal is a rarefaction pulse. The best-fit line to these results does not pass through zero at 4°C as expected. The zero crossing is 6.0° ± 0.2°C. (χ^2 for the linear fit is 14 for 11 DOF.) Systematic errors, such as thermometer calibration and temperature gradients in the acoustical medium, have been measured to be < 0.1°C. Corrections due to impurities in the water or microbubble production are naïvely expected to shift the zero-crossing down in temperature — not up as we observe. The origin of this discrepancy, the only significant deviation from thermal-model expectations, is not understood.

We have investigated those characteristics of the acoustic phenomena that could only exist if sound generation resulted from microbubble formation. The microbubble hypothesis suggests that the signal amplitude would be a function of the initial size and density of bubbles. The microbubble density is a function of the density of nucleation sites, so we expect the signal amplitude to be sensitive to the salt content and the amount of dissolved gas in the acoustic medium. We have investigated this by comparing samples of deionized water, water through which N_2 had been bubbled for one hour, and water in which NaCl was dissolved. No significant difference in signal amplitude was detected.

A further test for microbubble formation utilizes the dependence of the signal on bubble size. The average bubble size can be altered by changing the ambient pressure. If the ambient pressure approaches the internal bubble pressure at formation (calculated[5] to be ~ 1500 psi), the growth of the bubble should be inhibited and a smaller acoustic signal should be produced. Alternatively, a partial vacuum should increase the average microbubble radius and density (since the boiling point decreases), thereby intensifying the acoustic signal.

This test was performed with a vessel that could be pressurized to 1600 psi and alternatively evacuated to 17 mm Hg. Fig. 15, a graph of the resulting signal amplitude at various ambient pressures, reveals that the signal intensity is pressure independent. (The fit to a constant has $\chi^2 = 5.5$ for 6 DOF.)

In the linac experiment, the signal amplitude as a function of the hydrophone distance from the source (Fig. 16) demonstrates that the sonic pressure varies as 1/R, as anticipated in this far field configuration. (For a 4.5-cm by 30-cm deposition, $A = L^2/\lambda = 0.9$ m.)

The smallest observed signal in water at the cyclotron (corresponding to 2×10^{15} eV) was 0.2 dynes/cm^2, centered at 30 KHz (8 cm from the Bragg peak) and with a signal-to-noise ratio of 3:1. This establishes our signal-detection threshold under these conditions. Ultimately, the detection threshold is limited by thermal noise, which rises as $\sqrt{f}$ and has a value of 6×10^{-5} dynes/(cm^2 $\sqrt{\text{Hz}}$) at 30 KHz. In practice, this absolute minimum noise level is not easily attainable in the laboratory, due to both acoustical and electrical noise, particularly below 30 KHz. Figure 17 illustrates the noise spectrum in the acoustically isolated barrel of water used in the FEB experiment. In spite of external noise from water pumps, magnet hum, fans, etc., the noise above 30 KHz is primarily thermal (the theoretical minimum). The broadband features of the signal we wish to detect require opening the observation window to include most frequencies of interest. Typically, a center frequency of 30 KHz requires an observation window 30-KHz wide for minimum signal distortion. The average noise level in this window is ~3×10^{-4} dynes/(cm^2 $\sqrt{\text{Hz}}$). Thus, under average laboratory noise conditions in water at 20°C, we expect a wideband noise level of 0.05 dynes/cm^2.

Under these noise conditions (far from optimal), the signal-to-noise ratio is 1:1 for an energy deposition of 7×10^{14} eV. Optimal noise conditions suggest a detection threshold of 8×10^{13} eV for a single transducer.

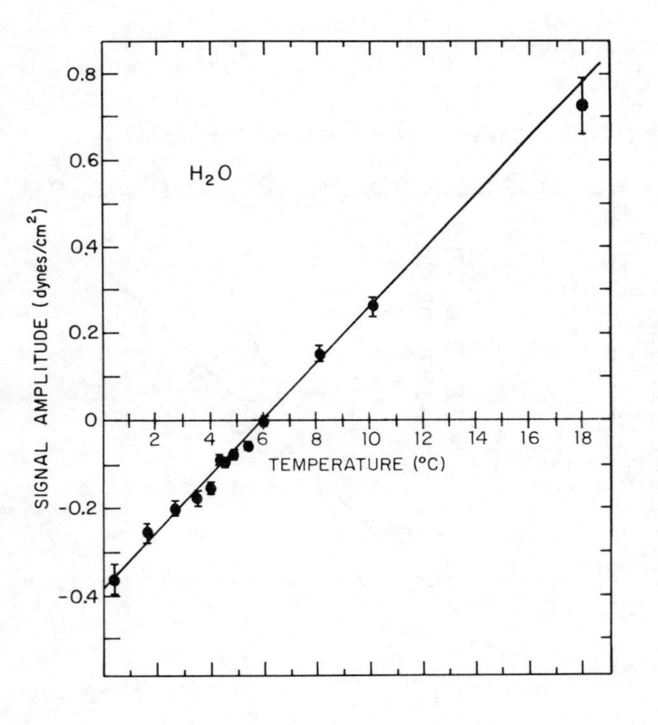

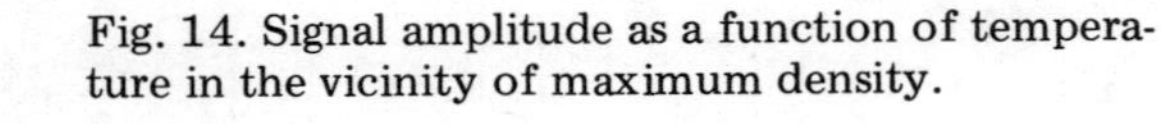

Fig. 14. Signal amplitude as a function of temperature in the vicinity of maximum density.

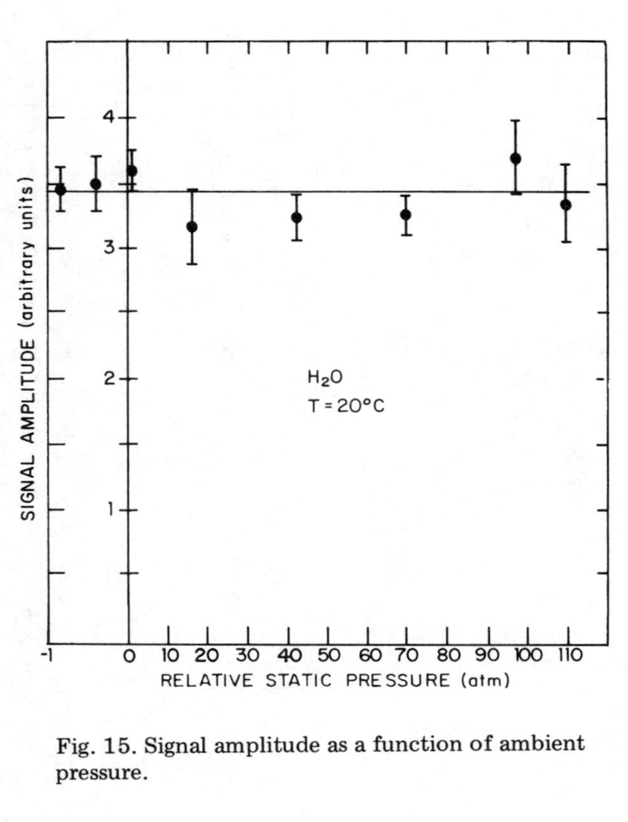

Fig. 15. Signal amplitude as a function of ambient pressure.

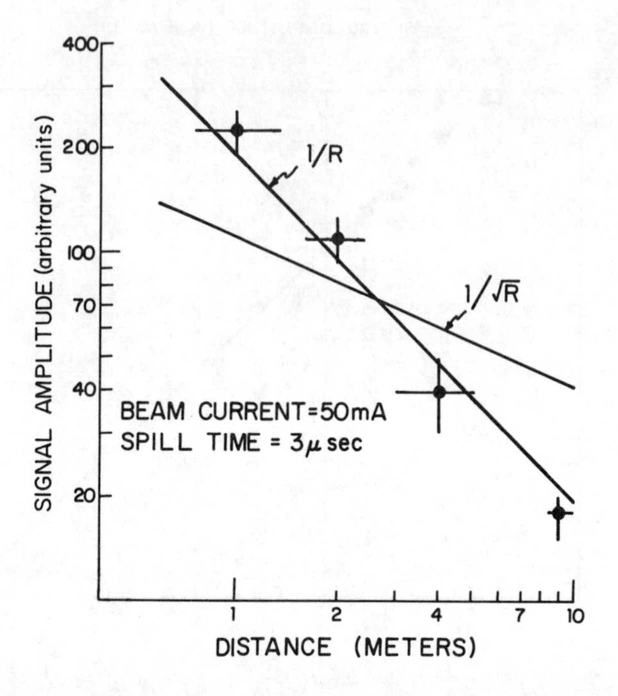

Fig. 16. Signal amplitude as a function of distance from the source.

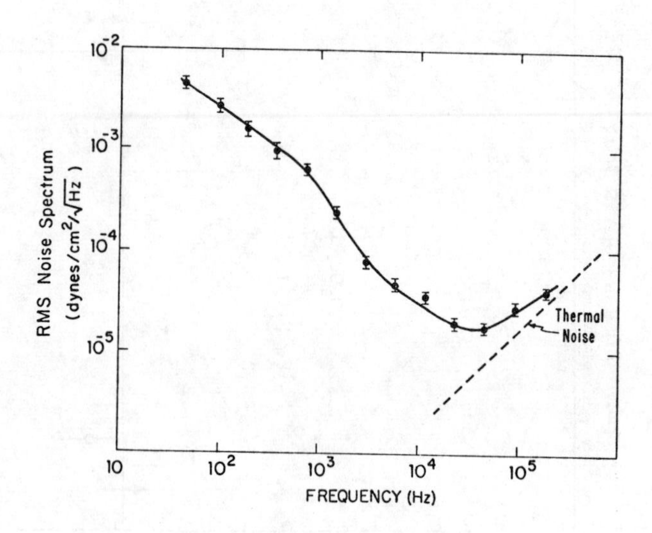

Fig. 17. Noise spectrum in the water barrel of the FEB experiment.

The properties of the acoustic signal produced by charged particles traversing liquids suggest many varied applications. Immediately, a new charged particle beam monitor is possible. For dE/dx energy losses $> 10^{14}$ eV, the signal is sufficient to activate several balanced transducers, which can be arranged to sense beam intensity, position, and diameter. For typical accelerator beams of about 10^{12} particles over a short spill duration, the 10^{14} eV threshold allows very thin detectors ($\sim 10^{-4}$ g); e.g., gas devices are clearly conceivable.

If the thermal mechanism dominates, highly charged heavy ion beams can be detected either with much less material in the beam (since the dE/dx losses are proportional to Z^2) or with lower thresholds. At high energies, as in cosmic rays, the Z^2 dependence of the sonic signals renders single heavy ions observable with low thresholds. A particular application[10] is the determination of the cosmic-ray isotope abundance for $Z > 30$, where no current detector makes an adequate measurement of Z. Here, gas-filled detectors, which leave the nucleus intact during the sonic measurement, are being developed.[10,11]

Calorimetry on the particle showers produced in high-energy physics experiments offers another possible area of application. For example, the next generation of neutrino experiments demands massive ($\gtrsim 10^4$ ton) detectors which, therefore, must be composed of inexpensive materials. Several properties of acoustic detectors are attractive: 1) the low cost of a single transducer ($\sim$ 100 times less than a photomultiplier tube); 2) the very large signal transmission length ($\sim$ 1 km vs 3 m for light in liquids and plastic), which reduces the sensing of a massive detector from a volume to a surface problem; 3) the savings allowed by the use of low-speed electronics required by sonic detectors; and 4) the wide range of sensitive materials (opaque materials such as Hg or even homogeneous solids are conceivable active detectors).

The Fourier transform of the signal should measure the proportion of energy deposited electromagnetically vs. hadronically. Very tight electromagnetic showers will populate high frequencies, whereas the 300 MeV/c characteristic transverse momentum of hadronic interactions produces wider showers and, therefore, lower frequencies. For hadronically induced showers, the fluctuations in the amount of energy shared between the π^0 and $\pi^\pm$ components early in the shower development limit the energy resolution in traditional scintillator and nonuranium liquid-argon calorimeters. (Due to binding energy loss, etc., the hadronic component transfers its energy into ionization much less efficiently.) In the sonic device, an appropriately heavier weighting of the fraction of the signal appearing at low frequencies (i.e., hadronically induced) should allow correction for this fluctuation.

The primary disadvantage of the sonic device is the high threshold for a single transducer, about 10^{14} eV in H_2O. To decrease the threshold in water to levels useable in the next generation of neutrino experiments (the neutrino energies at the Fermilab energy-doubler will be $\sim 10^{11}$ eV) would require about 10^4 phased transducers. Although this is a large number of sensors, the cost per transducer is small. On the other hand, substantially higher signal levels can be obtained through the use of materials of higher densities, temperatures, and K/C_p. For example, the value of K/C_p for liquid argon is several hundred times that for water.

Cosmic-ray neutrino detection is an area of high-energy physics requiring ultra-massive detectors. The very long sonic transmission lengths and the simple nature of the solid ceramic transducers are ideal for experiments which have been proposed[12] to instrument $\sim$ 1 to 100 km^3 ($\sim 10^9$ to 10^{11} tons) of homogeneous material. The use of fresh or salt water and of salt domes has been suggested. A distinct disadvantage of fresh water is that $K \rightarrow 0$ at 4°C, the temperature of water at great depths necessary for shielding these detectors from background. The value of K for salt water, on the other hand, is small but nonzero at 4°C (about 1/2 of that for fresh water at 20°C). At great depths in salt domes, salt retains its high K and large density, and becomes homogeneous.

Electromagnetic/hadronic-shower discrimination and hadronic-shower compensation are also important in these experiments. The high aspect ratio of showers at these energies ($> 10^{13}$ eV) imply a coherent sonic signature with a very small divergence ($\sim$ 6 mrad). Thus, the thin disk of sound emanating from the shower provides precise angular (as well as positional) information about the shower. This is important in locating extraterrestrial sources of events. The sign of the shower direction is also determined. Monte Carlo simulations of showers show that, even at these high energies, a distinctive spectral decrease occurs as the showers develop and get larger in diameter. Determining the sign of a shower's direction is fundamental to ascertain whether the neutrino traveled through the earth or not in reaching the detector.

The noise spectrum at great depths is anticipated to be dominated by thermal noise at high frequencies and by surface noise at low frequencies. For calm seas, the noise spectrum[13] shows a broad-band window at 30 KHz. Further, the very thin disk of sound emanating from the shower is a unique signature readily distinguishable from known backgrounds such as animals in the sea, seismic disturbances, etc.

We have demonstrated that an observable acoustic signal is produced in a single transducer by charged particle depositions $\gtrsim 10^{14}$ eV in fluid media. The source of the signal is dominantly thermal expansion. Applications to beam monitoring, heavy-ion experiments, and observing the showers produced by high-energy neutrinos at long distances from their source are foreseeable.

FOOTNOTES AND REFERENCES

1. Partial descriptions of the results reported here have appeared in two articles by T. Bowen et al., in Proceedings of the 1976 DUMAND Summer Workshop, Honolulu, 1976 (referred to below as DUMAND '76), edited by A. Roberts (Fermilab, Batavia, Illinois, 1977), p. 547 and p. 559; R. L. Sulak et al., in Proceedings of the Twelfth Rencontre de Moriond, Flaine, France, 1977, edited by J. Tran Than Van, p. 381; J. G. Blackington et al., J. Acoust. Soc. Am. 61 (S1), S80 (A), 1977; L. R. Sulak et al., in Neutrino '77, proceedings of the International Conference on Neutrino Physics and Neutrino Astrophysics, Baksan Valley, U.S.S.R., 1977, edited by M.A. Markov (Nauka, Moscow, 1978), Vol. 2, p. 350; T. Bowen et al., in Proceedings of the Fifteenth International Conference on Cosmic Rays, Plovdiv, 1977 (Bulgarian Academy of Science, Plovdiv, Bulgaria, 1977), Vol. 6, p. 270; and M. Levi et al., IIEE Trans. Nucl. Sci. NS-25, 325 (1978). A detailed description of the work has been submitted for publication in Nucl. Instrum. Methods.
2. B. L. Beron et al., IEEE Trans. Nucl. Sci. 17, 65 (1970); B. L. Beron and R. Hofstadter, Phys. Rev. Lett. 23, 184 (1969); R. M. White, J. Appl. Phys. 34, 3559 (1963); I. A. Borshkovsky et al., Zh. Eksp. Teor. Fiz. Pis'ma 13, 546 (1971) [JETP Lett. 13, 390 (1971)]; I. A. Borshkovsky et al., Lett. Nuovo Cimento 12, 638 (1975).
3. T. Bowen, DUMAND 76, p. 523. An updated version of this work appears in the Proceedings of the Fifteenth International Conference on Cosmic Rays, Plovdiv, Vol. 6, p. 277.
4. DUMAND 76, p. 553.
5. G. A. Askarjan and B. A. Dolgoshein, "Acoustic Detection of High Energy Neutrinos at Big Depths," preprint No. 160, Lebedev Institute of the USSR. Academy of Sciences, Moscow, 1976.
6. V. D. Volovik et al., "Acoustic Radiation Produced during Molecular Dissociation," in Proceedings of the Ninth USSR Acoustics Conference, Moscow, 1977.
7. The beam line and tank are part of the Brookhaven Linac Isotope Producer. Fred L. Horn has described the facility in the Proceedings of the Twentieth Conference on Remote Systems Technology, Idaho Falls, Idaho, 1972, edited by R. Farmakes (Am. Nuc. Soc., Hinsdale, Illinois, 1972), p. 310.
8. Handbook of Chemistry and Physics (Chemical Rubber Co., Cleveland, 1975), 56th ed.; F. Kaye and T. Laby, Tables of Physical and Chemical Constants (Longmans, Green, London, 1957), 11th ed.
9. This graph was derived from data which appears in M. W. Zemanski, Heat and Thermodynamics (McGraw-Hill, New York, 1957), 4th ed., p. 259.
10. W. V. Jones, in Proceedings of the Fifteenth International Conference on Cosmic Rays, Plovdiv, Vol. 6, p. 271.
11. G. Yodh (private communication).
12. The capabilities and design of a deep underwater muon and neutrino detector (DUMAND) have been explored in depth during several summer studies. The work of the most comprehensive study appears in DUMAND 76 (see Footnote 1).
13. E. M. Arase and T. Arase, in Acoustics and Vibration Progress, (Wiley, New York, 1974), Vol. 1, p. 200.

NEUTRINO OSCILLATIONS

A. K. Mann
Department of Physics, University of Pennsylvania,* Philadelphia, Pennsylvania 19104

ABSTRACT

A summary is presented of various possible and active experiments to search for neutrino oscillations.

INTRODUCTION

Since the initial suggestion of Pontecorvo[1] that neutrinos of different types may exhibit oscillations of type as a function of time, similar to the oscillations between K^0 and $\overline{K}^0$, it has been of interest to consider possible experimental searches for neutrino oscillations. An experiment which observed neutrino oscillations would lead immediately to three important results: (i) that the mass of at least one of the neutrino types is nonzero, (ii) that the separate conservation of muon lepton-number and electron lepton-number does not hold, and (iii) a determination of the total number of neutrino types (i.e., ν_μ, ν_e, . . .) with nondegenerate masses that are mixtures of mass eigenstates. Items (i) and (ii) are independent, necessary conditions for the existence of neutrino oscillations in vacuum. Neutrino oscillations may also develop, even if all neutrinos have zero mass, as a result of coherent scattering through the weak neutral current when passing through matter, provided that current is off-diagonal in neutrino type.[2]

The theory of neutrino oscillations has been developed in detail in several papers.[3] For simplicity, we confine ourselves here to the case of two neutrino types; extension to more than two neutrino types can be made directly. Briefly, the wavelength of the oscillations ℓ_{osc} is given by

$$\ell_{osc} = \frac{(4\pi\hbar/c^2)p_\nu}{[m(\nu_2)]^2 - [m(\nu_1)]^2}, \tag{1}$$

where p_ν is the momentum of the initial neutrinos and $m(\nu_2)$ and $m(\nu_1)$, which are linear combinations of $m(\nu_\mu)$ and $m(\nu_e)$, are the masses of the eigenstates ν_1 and ν_2 of the world hamiltonian that includes a part $H^{(1)}$ responsible for the neutrino oscillations.

Starting with a neutrino of type ν_μ at t = 0, the probability of finding a neutrino of type ν_e at time t as the result of neutrino oscillations is

$$P(\nu_e;t|\nu_\mu;0) = \frac{1}{2}\left[\frac{m_{e\mu}^2}{[m(\nu_\mu) - m(\nu_e)]^2 + m_{e\mu}^2}\right][1 - \cos(E_2 - E_1)t], \tag{2}$$

where $m_{e\mu} = 2\mathrm{Re} < \nu_e | H^{(1)} | \nu_\mu >_{p_\nu = 0}$ is the off-diagonal matrix element responsible for the $\nu_\mu \leftrightarrow \nu_e$ mixing and E_2 and E_1 are the energies of ν_1 and ν_2.

ISSN:0094-243X/79/520101-07$1.50

*Research aided in part by the U.S. Department of Energy.

In the simplest, perhaps most probable, case $m(\nu_\mu) = m(\nu_e)$ so that $m(\nu_1) = m(\nu_\mu) - m_{e\mu}/2$ and $m(\nu_2) = m(\nu_\mu) + m_{e\mu}/2$, yielding

$$P(\nu_e;t|\nu_\mu;0) = 1/2\left[1 - \cos(E_2 - E_1)t\right], \tag{3}$$

$$\langle P(\nu_e;t|\nu_\mu;0)\rangle_{\substack{\text{time}\\ \text{average}}} = 1/2, \tag{4}$$

and

$$E_2 - E_1 \cong \frac{1}{2p_\nu}\left\{[m(\nu_2)]^2 - [m(\nu_1)]^2\right\} = \frac{1}{2p_\nu}\,[2m(\nu_\mu)]\,m_{e\mu}, \tag{5}$$

from which the expression above for ℓ_{osc} follows, because

$$\ell_{osc} = \frac{2\pi\hbar c}{E_2 - E_1}. \tag{6}$$

The expressions for ℓ_{osc} and $P(\nu_e;t|\nu_\mu;0)$ provide the basic criteria for evaluation of a neutrino oscillation experiment. (1) The smaller the value of the unknown quantity $[m(\nu_2)]^2 - [m(\nu_1)]^2 \equiv [m(\nu_2) + m(\nu_1)]\,[m(\nu_2) - m(\nu_1)]$ that can be reached in an experiment, the better the experiment. From Eq. (1),

$$M^2 \equiv [m(\nu_2)]^2 - [m(\nu_1)]^2 \propto \frac{p_\nu}{\ell_{osc}},$$

so that the lowest value of p_ν combined with the largest value of the neutrino source-detector distance (x) makes for the most sensitive experiment. For a given neutrino source-detector distance, the most sensitive oscillation experiment is obtained with neutrinos of the lowest p_ν compatible with sufficient count rate in the detector to permit a statistically significant experiment to be done. (2) Since it is desirable to search for neutrino oscillations in the region of low probability, i.e., values of $P(\nu_e;t|\nu_\mu;0)$ smaller than in the simplest (maximal oscillation) case [Eq. (3)] which yields the largest time-averaged probability [Eq. (4)], it is also necessary to minimize the background of spurious counts in the detector that may simulate neutrino oscillations.

POSSIBLE EXPERIMENTS

A. General

It is of interest to consider briefly in a general way the various types of searches for neutrino oscillations that are at present feasible. These depend on different combinations of source-detector distance (x), initial neutrino momentum (p_ν), and on means of minimizing backgrounds. A list of possible experiments is given in Table I, which shows the figure of merit $M \equiv \left\{[m(\nu_2)]^2 - [m(\nu_1)]^2\right\}^{1/2}$, the combination of x and p_ν, and the experimental signal of neutrino oscillations for each type of experiment. It is clear from Table I that the nature of a neutrino oscillation experiment is determined primarily by the neutrino source.

(a) The searches for oscillations using very low energy $\bar{\nu}_e$ from reactors are limited to relatively small source-detector distances by the low antineutrino interaction cross section at those energies. In addition to the primary interaction, $\bar{\nu}_e + p \rightarrow e^+ + n$ (the reaction observed by Cowan and Reines in their original experiment to detect antineutrinos), there are weak-neutral-current channels $\bar{\nu}_e + p \rightarrow \bar{\nu}_e + p$ and $\bar{\nu}_e + n \rightarrow \bar{\nu}_e + n$,

Table I. Possible neutrino oscillation experiments.

Source	E_ν (MeV)	x (km)	M_{min} (eV)	Type
Reactor	$1-10$	$10^{-2}-10^{-1}$	0.05	$\bar{\nu}_e + p \rightarrow e^+ + n$, inverse square law test
Meson factory	$10-10^2$	$10^{-1}-1$	0.2	$\nu_e + n \rightarrow e^- + p$, proof of reaction[a]
High-energy accel.	10^2-10^4	$1-10^3$	0.05	as in meson factory, or ν_e/ν_μ at x_1, x_2
Sun	$0.2-10$	10^8	10^{-6}	absolute measurement; compare with solar-model calculation

[a]See discussion in text.

and possibly background reactions from a contamination of ν_e in the initial beam. The signal for neutrino oscillations in such experiments is a departure from the inverse-square law, and perhaps a relative measurement obtained by moving the detector a short distance from one position (x_1) to another (x_2). It is not yet clear what the sensitivity of reactor experiments is to values of, say, $P(\bar{\nu}_\mu;t|\bar{\nu}_e;0)$ far from the maximum value.

In a recent report by Sobel and Reines,[4] it is claimed that using data from an earlier (1966) reactor antineutrino experiment they obtain $\{[m(\nu_2)]^2 - [m(\nu_1)]^2\}^{1/2} \lesssim 0.5$ eV/c^2 for the maximal oscillation case by means of an inverse square law test. In that experiment p_ν was about 3.3 MeV/c and the source-detector distance 6 m. A second, more recent experiment carried out at the same reactor at 11.1 m is being analyzed, and a third experiment with a high power reactor (San Onofre), now under consideration, will allow source-detector distances between 25 and 100 meters.

A suggestion for a similar experiment at the Grenoble (France) reactor has been made by Boehm and Mössbauer,[5] who propose both an inverse-square-law test of comparable sensitivity and a test in which the detector is moved from one position to another.

(b) Experiments utilizing somewhat higher energy neutrinos, to obtain higher cross sections, are possible at medium and high-energy proton accelerators. With these accelerators, a different class of neutrino oscillation experiments becomes possible. Since the neutrinos produced are predominantly of the ν_μ type, the observation of events uniquely attributable to ν_e interactions in a ν_μ beam would, in principle, be a relatively clear, sensitive signature of neutrino oscillations. Hence, below the threshold for quasielastic ν_μ scattering, the neutrino oscillation experiment becomes a search for quasielastic ν_e scattering rather than an inverse-square-law test as at lower energies. This nominally attractive possibility requires detailed experimental testing before its feasibility can be determined.

We mention a special case as illustrative of an accelerator experiment with high sensitivity using ν_μ with energies below the quasielastic scattering threshold. To eliminate the background from neutral-current ν_μ interactions and indeed from almost all other background sources, it is tempting to make use of the inverse of the well known beta-decay process

$$\nu_e + {}^{12}C \rightarrow e^- + {}^{12}N^*$$

followed by

$${}^{12}N^* \rightarrow {}^{12}C + e^+ \quad (\langle E_e \rangle \approx 8 \text{ MeV, } 13 \text{ msec})$$

as the signature of ν_e interaction, rather than quasielastic ν_e scattering.[6] The cross sections for the reactions $\nu_e + {}^{12}C \rightarrow e^- + {}^{12}N^*$ and $\nu_\mu + {}^{12}C \rightarrow \mu^- + {}^{12}N^*$ have been calculated as a function of incident energy by O'Connell, Donnelly, and Walecka.[7] For the positive-parity excited state in ${}^{12}N$ which is of interest here, the cross section for electron production at 100 MeV $\approx 0.5 \times 10^{-39}$ cm^2/nucleus should be reduced by a factor of 2 to agree with electron-scattering data, according to the authors, and by another factor of 6 to convert into the same units as the quasielastic cross section for $\nu_e + n \rightarrow e^- + p$. One obtains $\sigma(\nu_e + {}^{12}C \rightarrow e^- + {}^{12}N^*$ at 100 MeV) $\approx 4 \times 10^{-41}$ cm^2/neutron compared with $\sigma(\nu_e + n \rightarrow e^- + p$ at 100 MeV) $\approx 1 \times 10^{-39}$ cm^2/neutron, which indicates, unfortunately, that the cleanliness of this method is achieved at a considrable sacrifice in overall rate. In addition, this or the previous method would require a convincing empirical demonstration that the energy of all ν_μ in the incident beam did not exceed the ν_μ quasielastic scattering threshold.

(c) A small further increase of the energy of the incident ν_μ beam leads to still another class of experiments in which both ν_μ and ν_e interactions are detected. Observe that the experimental method is different from those employed at the lower energies, since at energies greater than the quasielastic scattering threshold it is the ratio of all recognized ν_e-induced events to all recognized ν_μ-induced events that is measured. The advantage of having events of the type $\nu_\mu + n \rightarrow \mu^- p$ is to normalize the experiment empirically. It is desirable to initiate a neutrino oscillation study above the threshold for ν_μ quasielastic scattering, where the experiment is most direct. Tests could be made to justify a subsequent experiment below threshold and ultimately each experiment could be used as a check on the other.

Finally, the energy of the incident ν_μ beam may be raised even more to take advantage of the linear dependence of the total ν_μ-nucleon cross section on E_ν. An appropriate trade-off between count rate and source-detector distance allows an experiment of high sensitivity to be done at the level $M \lesssim 0.1$ eV/c^2, over a distance which is a significant fraction of the earth's radius.[3]

(d) As is evident from Table I, an oscillation experiment using the sun as the neutrino source is, in principle, capable of great sensitivity. The limitation is, however, the absence of an empirical means of normalizing the experiment. At present, the rate of neutrino interactions observed on earth[8] (presumably) from the sun is compared with an expected rate calculated from an elaborate solar model.[9] The experimental value[8] is less by a factor of about 3 than the calculated value,[9] where the principal uncertainty in the latter result arises from the measured nuclear-physics cross sections that are put into the solar-model calculation. If the experimental and calculated rates are taken at face value, they suggest a discrepancy which might be resolved by the existence of neutrino oscillations. But it is too early to consider this comparison as more than suggestive.

It has been suggested[10] that a procedure for normalizing neutrino rates from the sun might be provided by the requirement of self-consistency among the measurements of neutrino rates from different neutrino-producing reactions in the sun. Thus, observation on earth of neutrinos from the main thermonuclear reaction $p + p \rightarrow d + e^+ + \nu_e$ (and from $e^- + p + p \rightarrow d + \nu_e$) might be used in conjunction with the reaction already observed[8] to estimate in a reliable way the number of neutrinos expected to reach the earth.

One salient conclusion to be drawn from this description of possible neutrino oscillation experiments is that, although the level of sensitivity measured by the parameter $\{[m(\nu_2)]^2 - [m(\nu_1)]^2\}^{1/2}$ is roughly similar for all terrestrial experiments, the systematic uncertainties differ significantly. Thus, it is of value to carry out searches for neutrino oscillations by all of the different methods, the limitations being set by feasibility and relative expense. It is clearly desirable to push ahead with further studies of solar neutrinos.

B. Specific.

In addition to neutrino-oscillation limits from reactor data and from the sun, there are data available from one accelerator experiment, and more such experiments are planned or in progress.

An accelerator experiment was carried out at CERN by Bellotti et al.,[11] who used a heavy liquid bubble chamber as a detector and a wide-band neutrino beam of average momentum 2.26 GeV/c. The source-detector distance was approximately 65 m. The experiment yielded the limit $M \leqslant 1\ \mathrm{eV/c^2}$ for complete $\nu_\mu - \nu_e$ mixing, i.e., the maximal oscillation case.

A test for a neutrino oscillation experiment was also done at BNL.[12] Here the source-detector distance was 165 m, and the detector was a multi-ton liquid scintillator calorimeter.[13] The synchrotron at BNL was operated to impinge protons of momentum 1.5 GeV/c on the neutrino-producing target. This gave rise to a neutrino event rate in the detector that was maximized in the energy region between 150 and 250 GeV. An experiment based on these values of the parameters is expected to yield a limit on M of a few-tenths of an $\mathrm{eV/c^2}$ for complete mixing.

An accelerator experiment involving a long source-detector distance ($\gtrsim 10^3$ km) has also been suggested.[3] To achieve such a long path for the neutrinos requires that the accelerator produced neutrino beam be directed into the earth in an arrangement similar to that shown in Fig. 1. A suggested geographical arrangement that would make use of the proton synchrotron at FNAL (Fermi National Accelerator Laboratory) is shown in Fig. 2. This experiment (which is described in more detail in reference 3) might achieve a limit on M of about 0.1 $\mathrm{eV/c^2}$, and would also test for as yet unobserved interactions of neutrinos with matter.[2]

CONCLUSION

In this brief survey of neutrino oscillation experiments I have tried to indicate that a wide variety of such experiments is possible. The neutrino sources and the signatures for neutrino oscillations are very different for different types of experiments. This justifies a multiplicity of searches to ensure that neutrino oscillations are not missed by the limitations of a given type of experiment. It is probable that in the next few years several terrestrial experiments of increasing sensitivity will be carried out, some with neutrinos traversing long paths in matter, as an effort is made to tighten the quantitative limit on the existence of neutrino oscillations. Neutrinos from the sun, which also will

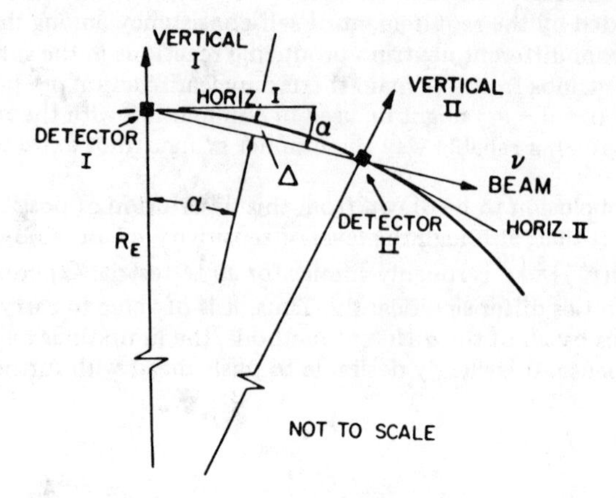

Fig. 1. Geometry of a feasible experiment. If the distance between detectors I and II is 1000 km, then $\alpha = 0.078$ rad and $\Delta = 19$ km. R_E is the radius of the earth = 6.4×10^3 km.

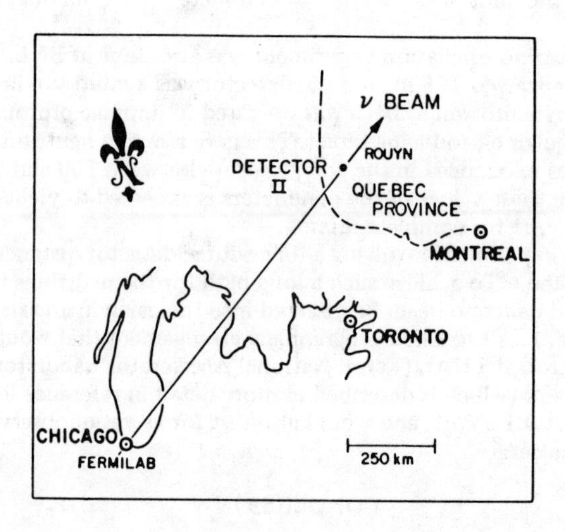

Fig. 2. Approximate geography of the proposed experiment. The present neutrino beam at Fermilab is directed 38° 13' 52" east of north as indicated roughly.

be studied in greater detail in the next decade,[14] offer the possibility of a very sensitive search for neutrino oscillations, but a conclusive answer will depend on how compelling can be made the determination of the expected neutrino rate on earth based on the current solar model.

REFERENCES

1. B. Pontecorvo, Zh, Eksp. Teor. Fiz. 53, 1717 (1967) [Sov. Phys. — JETP 26, 984 (1968)].
2. L. Wolfenstein, Phys, Rev. D 7, 2369 (1978).
3. See, e.g., A. K. Mann and H. Primakoff, Phys, Rev. D 15, 655 (1977), and references therein.
4. H. W. Sobel and F. Reines, UC-Irvine preprint (unpublished).
5. F. Boehm (private communication).
6. An experiment based on this idea was proposed at BNL by E. Egelman et al.
7. J. S. O'Connell, T. W. Donnelly and J. D. Walecka, Phys. Rev. C 6, 719 (1972); H. Überall, B. A. Lamers, J. B. Langworthy, and F. J. Kelly, Phys. Rev. C 6, 1911 (1972).
8. J. K. Rowley, B. T. Cleveland, R. Davis, Jr., and J. C. Evans, in Neutrino '77, proceedings of the International Conference on Neutrino Physics and Neutrino Astrophysics, Baksan Valley, U.S.S.R., 1977, edited by M. A. Markov (Nauka, Moscow, 1978), Vol. 1, p. 15; BNL Report No. BNL 23418 (unpublished).
9. J. N. Bahcall, Astrophys. J. Lett. 216, L115 (1977).
10. S. M. Bilenky and B. Pontecorvo, Nuovo Cimento Lett. 17, 569 (1976); H. Primakoff, in Proceedings of the Informal Conference on Status and Future of Solar Neutrino Research, edited by G. Friedlander (BNL Report No. BNL 50879), Vol. I, p. 211.
11. E. Bellotti et al., Nuovo Cimento Lett. 17, 553 (1976).
12. This was a test of type (c) carried out by A. Entenberg, W. Kozanecki, J. Horstkotte, A. K. Mann, C. Rubbia, J. Strait, L. Sulak, P. Wanderer, and H. H. Williams. See L. R. Sulak et al., in Neutrino '77, Vol. 2, p. 280.
13. D. Cline et al., Phys. Rev. Lett. 37, 252 and 648 (1976).
14. See, e.g., Ref. 8 and also G. T. Zatsepin, in Neutrino '77, Vol. 1, p. 20.

EFFECTS OF MATTER ON NEUTRINO OSCILLATIONS

L. Wolfenstein
Carnegie-Mellon University, Pittsburgh, Pennsylvania 15213

ABSTRACT

Coherent forward scattering of neutrinos from atoms can affect neutrino oscillations for path lengths of the order of 10^8 cm or more of terrestrial matter. Two possibilities are discussed: (1) matter-induced oscillations for massless neutrinos, (2) the modification of vacuum oscillations when the vacuum path is replaced by matter.

The usual discussion of neutrino oscillations is based upon the original idea of Pontecorvo[1] of vacuum oscillations. Some proposed experiments[2,3] involving high-energy neutrinos, however, involve the passage of the neutrinos through a large distance of terrestrial matter. It has recently been shown[4] that for distances of the order of 10^8 cm or more neutrino oscillations in matter may be quite different from oscillations in vacuum.

The basic idea is that the index of refraction of neutrinos in matter can produce a significant change of phase. For example, for neutrinos interacting with electrons via the effective weak interaction

$$\frac{G}{\sqrt{2}}\bar{\nu}\gamma_\lambda(1+\gamma_5)\nu\bar{e}\gamma_\lambda e, \tag{1}$$

the index of refraction is given by

$$k(n-1) = \frac{2\pi N_e}{k} f(0) = GN_e, \tag{2}$$

where N_e is the number of electrons per unit volume, G is the Fermi constant, and f(0) is the weak forward elastic-scattering amplitude of neutrinos. The characteristic length for a change of phase of 2π is

$$\ell_0 \equiv \frac{2\pi}{k(n-1)} = \frac{2.7 \times 10^9\,\text{cm}}{\rho_e}, \tag{3}$$

where ρ_e is N_e divided by 6×10^{23} cm^{-3}.

We consider two theoretical contexts in which this basic idea is applied. (1) For massless neutrinos, for which vacuum oscillations are impossible, matter oscillations can occur in a theory in which the neutral current changes neutrinos from one type to another. (2) If neutrinos are massive and vacuum oscillations do occur, these oscillations are described by different equations than in vacuum. This is true when ν_e are involved even if the neutral currents are the standard diagonal ones.

We first consider the case of massless neutrinos with an effective neutral-current interaction given by[5]

ISSN:0094-243X/79/520108-05$1.50

$$H_w = \frac{G}{\sqrt{2}} J_\lambda L_\lambda ,$$

$$L_\lambda = \cos\alpha[\overline{\nu}_\mu \bullet \nu_\mu + \overline{\nu}_e \bullet \nu_e + \overline{\nu}_\tau \bullet \nu_\tau]$$
$$+ \sin\alpha[\overline{\nu}_\mu \bullet (\nu_\tau \cos\beta + \nu_e \sin\beta)$$
$$+ \overline{\nu}_\tau \bullet \nu_e \sin\gamma + \text{h.c.}] , \tag{4}$$

where the dot • stands for $\gamma_\lambda(1+\gamma_5)$, and where ν_e, ν_μ, and ν_τ are the three types of left-handed neutrinos and J_λ is the neutral current for particles other than neutrinos. The normal theoretical neutral-current interaction has $\alpha = 0$; however, it is impossible from present experiments[6] to tell whether in neutral-current interactions ν_μ stays as ν_μ or changes into ν_e or ν_τ. Thus, for a given J_λ the experimental results for ν_μ are independent of the choices of α, β, and γ. The piece of the current J_λ that is of interest here is the <u>vector</u> current for protons, neutrons, and electrons

$$J_\lambda = g_p \overline{p}\gamma_\lambda p + g_n \overline{n}\gamma_\lambda n + g_e \overline{e}\gamma_\lambda e , \tag{5}$$

since this contributes to the coherent forward-scattering amplitude f(0) from atoms. The quantities g_p and g_n are fairly well determined[7] from ν_μ interactions with nuclei but there is only fragmentary information on g_e; in our numerical examples we choose[8] $g_e = -0.5$. When $\alpha \neq 0$, the eigenstates for propagation through matter are not the states ν_e, ν_μ, and ν_τ, which are the states defined by the charged-current interaction. As a result, a beam which is originally ν_e or ν_μ, as determined by its source, will oscillate into the other types as it passes through matter. This is analogous to the phenomenon of optical birefringence, in which case two planes of polarization are eigenvectors and beams with other states of polarization are transformed as they pass through the crystal.

For simplicity, we discuss here examples in which the oscillations involve only two states. In this case, a beam originally ν_μ has the probability of remaining ν_μ after going a distance x given by

$$I \equiv |\langle\nu_\mu|\nu_\mu(x)\rangle|^2 = 1 - \tfrac{1}{2}\sin^2 2\theta.[1 - \cos\Delta] ,$$

$$\Delta = \frac{2\pi}{\ell}\int_0^x \rho_e(x')\,dx' , \tag{6}$$

where θ is the mixing angle and ℓ the oscillation length in matter with $\rho_e = 1$ (typically, this means a density of 2 gm/cc). Two special cases are the following:

A. $\beta = \gamma = 0$; $\theta = \frac{\pi}{4}$; ν_μ oscillates only into ν_τ.

$$\ell = \frac{2.7 \times 10^9 \text{ cm}}{(g_p + g_e + y\, g_n)\sin\alpha} , \tag{7}$$

where $y = N/Z$ is the neutron-proton ratio.

B. $\beta = \frac{\pi}{2}$, $\gamma = 0$; ν_μ oscillates only into ν_e. In the case when ν_e is involved, one must take into account not only the neutral current but also the forward scattering of ν_e from electrons due to the charged-current interaction, which after a Fierz transformation can be written as Eq. (1). We then find that

$$\tan 2\theta = 2\sin\alpha(g_p + g_e + y g_n) , \tag{8a}$$

$$\ell = (2.7\times 10^9 \text{cm})\cos\theta . \tag{8b}$$

We now apply this idea to the proposed experiment[3] to observe the angular dependence of the intensity of cosmic-ray muon-type neutrinos coming through the earth.[9] Some results are shown in Fig. 1, where I is the transmission probability from Eq. (6) and ψ is the angle measured from the vertical. For each example,[10] we have set $y = 1$ and have assumed[7,8] $g_p + g_n + g_e = -1.2$. The case $\sin\alpha = 0.5$ implies that one-quarter of the neutral-current interactions are off-diagonal. For oscillations into either ν_e or ν_τ, such a large value of $\sin\alpha$ results in dramatic changes in the neutrino intensity as a function of the neutrino direction ψ. Therefore, this experiment would provide a sensitive test as to whether ν_μ changes to another type of neutrino in neutral-current interactions; I know of no other practical way to test this. How sensitive the test is depends on the value of $g_p + g_e + g_n$; as noted above this can be determined from neutral-current data, except that so far g_e is very uncertain and, in any case, the relative sign of g_e and $(g_p + g_n)$ cannot be determined. For oscillations into ν_τ (case A), even such a small value as $\sin\alpha = 0.1$ could produce the large effect seen in Fig. 1. It is conceivable that values of $\sin\alpha$ at this level might occur in some gauge models.[4] I have not shown the case of oscillations into ν_e (case B) for $\sin\alpha = 0.1$, since then the amplitude of the oscillations is very small (minimum value of I is about 0.95).

We now shift gears and go back to the original idea[1] of neutrino oscillations. This means we assume that neutrinos are massive and that the mass eigenstates are mixtures of ν_e, ν_μ, and ν_τ. At the same time, we will assume that the neutral current is normal; that is, $\alpha = 0$ in Eq. (4). The experiments we have discussed can distinguish this original idea from the case of oscillations induced by matter in several ways, as shown in Fig. 1:

1) The oscillations induced by matter have an oscillation length constrained by Eqs. (7) or (8).

2) The oscillation pattern as a function of ψ for vacuum oscillations from ν_μ to ν_τ depends only on the neutrino path length, whereas for matter-induced oscillations the pattern depends also on the density of the traversed matter, which is larger for small ψ when the neutrinos pass through the core of the earth.

3) For matter-induced oscillations, there is no dependence on neutrino momentum, whereas for the vacuum oscillation mechanism involving ν_μ going to ν_τ the oscillation length is directly proportional to the momentum.

For the case of the vacuum oscillation mechanism for ν_μ going to ν_e, the oscillation pattern is affected by the matter, provided the vacuum oscillation length is comparable to ℓ_0 defined by Eq. (3). This results from the fact that the charged-current interaction of ν_e with atomic electrons, which can be written in the form of Eq. (1), changes the phase of the ν_e component relative to ν_μ. As a result, the dependence on ψ and on momentum of these oscillations for ν_μ to ν_e differs from the case for ν_μ to ν_τ. To obtain the transmission probability I, it is necessary to solve the coupled differential equations

$$i\frac{d}{dx}\begin{bmatrix} a_e \\ a_\mu \end{bmatrix} = -\frac{\pi}{\ell_v}\begin{bmatrix} \cos 2\theta_v - \frac{2\ell_v}{\ell_0} & \sin 2\theta_v \\ \sin 2\theta_v & -\cos 2\theta_v \end{bmatrix}\begin{bmatrix} a_e \\ a_\mu \end{bmatrix}, \tag{9}$$

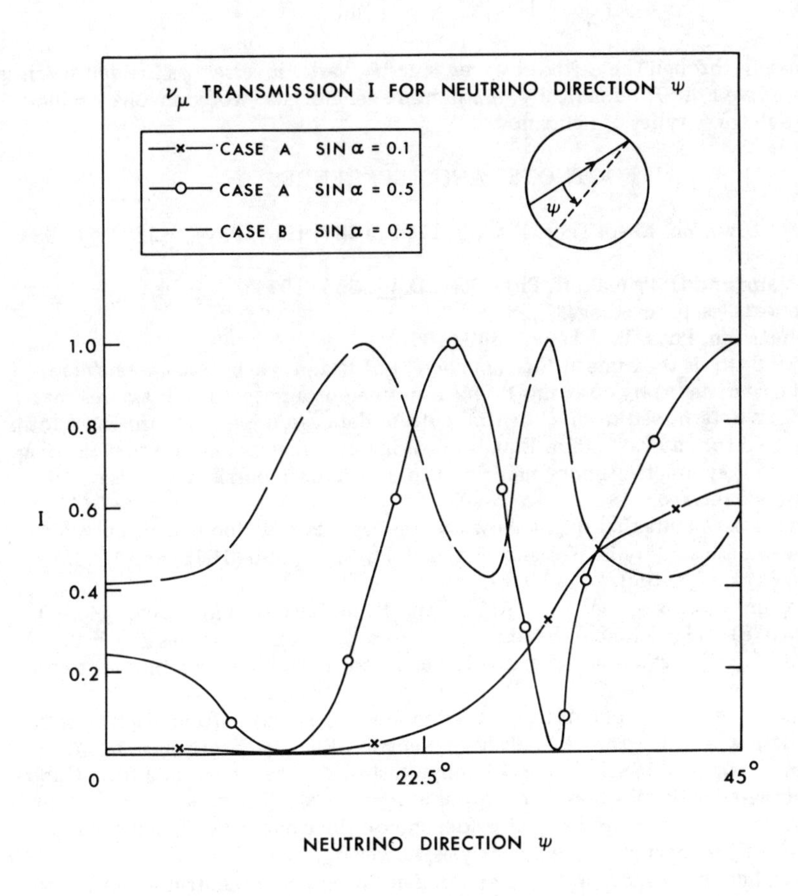

Fig. 1. Angular dependence of the transmission probability I of cosmic-ray muon-type neutrinos coming through the earth.

where $a_e(a_\mu)$ are the amplitudes of $\nu_e(\nu_\mu)$ and ℓ_v, θ_v are the vacuum oscillation length and mixing angle. For values $|\ell_v| \ll 10^9$ cm, the term ℓ_v/ℓ_0 can be neglected and the results are the same as for the vacuum case. For values $|\ell_v| \gg 10^9$ cm, the oscillation amplitude is small even if θ_v is 45°. For the general case, the solution is simple for a homogeneous medium:

$$I = 1 - \frac{1}{2}\left[\frac{\ell_m}{\ell_v}\right]^2 \sin^2 2\theta_v \left[1 - \cos\frac{2\pi x}{\ell_m}\right],$$

$$\ell_m = \ell_v \left\{1 + \left[\frac{\ell_v}{\ell_0}\right]^2 - 2\left[\frac{\ell_v}{\ell_0}\right] \cos 2\theta_v\right\}^{-1/2}.$$

For transmission through the earth, however, since ℓ_0 varies inversely as the density, it is necessary to solve Eq. (9) numerically. A still more general case would involve oscillations among all three types of neutrinos.

FOOTNOTES AND REFERENCES

1. B. Pontecorvo, Zh. Eksp. Teor. Fiz. 53, 1771 (1967) [Sov. Phys. — JETP 26, 984 (1968)].
2. A. K. Mann and H. Primakoff, Phys. Rev. D 15, 655 (1977).
3. K. Lande, these proceedings.
4. L. Wolfenstein, Phys. Rev. D 17, 2369 (1978).
5. With $\alpha = 0$ this is the same as Eqs. (1) and (2) of Ref. 4, except that we have replaced $\cos^2\alpha(\sin^2\alpha)$ by $\cos\alpha(\sin\alpha)$. This is more appropriate, since it assures that when J_λ is determined from ν_μ neutral current-data assuming $\alpha = 0$, the same form can be used for any value of α. Equation (4) involves the assumption that the diagonal piece is symmetric among neutrino types; still more possibilities arise if this condition is relaxed.
6. Possible experiments that might show up nondiagonal neutrino neutral currents have been discussed by L. Wolfenstein, Nucl. Phys. B91, 95 (1975) and L. M. Sehgal, Phys. Lett. 55B, 205 (1975).
7. For a recent discussion, see L. F. Abbott and R. M. Barnett, Phys. Rev. Lett. 40, 1303 (1978) and references therein. Here we use their central values $g_p = -0.1$, $g_n = -0.6$ for our examples; there still remains considerable uncertainty as to the correct values. The overall sign of (g_p, g_n) is not determined.
8. The value $g_e = -0.5$ is one of the two values derived from neutrino-electron scattering data; see H. Faissner, "Analysis of Neutrino-Electron Scattering Data", Aachen Report No. ISSN 0340-8701 (unpublished). More recent data from Gargamelle disagree with this, however. Insofar as $\bar{\nu}_e - e$ scattering data are used in the analysis, the value derived for g_e does depend on the values of α, β, and γ; present data do not justify such a refined analysis, however.
9. Some applications to the proposed experiment using FNAL neutrinos[2] and to the solar neutrino problem have been discussed previously.[4]
10. Numerical results are based on the density variation for the earth given in Table 1.3 of J. A. Jacobs, The Earth's Core (Academic, New York, 1975).

TELECOMMUNICATION BY HIGH-ENERGY NEUTRINO BEAMS*

F. J. Kelly, A. W. Sáenz, and H. Überall**
Naval Research Laboratory, Washington, D.C. 20375

"Give reign to many-throated fancy,
to reason, thought . . . but, mark it well!
do not forget extravagancy!"

Goethe, Faust I: Prologue in the Theater

ABSTRACT

High-energy neutrino beams, being able to penetrate the entire earth, may offer some unconventional possibilities for telecommunications on a global scale.

INTRODUCTION

In this paper, we consider the potential use of multi-GeV muon neutrino beams from high-energy accelerators, such as the one at Fermilab, for telecommunication purposes.

Two of the present authors had the privilege of knowing Prof. C. L. Cowan during his lifetime, one (F.J.K.) as a graduate student at Catholic University and the other (H.Ü.) as a colleague and collaborator at that institution. Their collaboration was a fruitful one, resulting in the first published suggestion[1] for employing the Čerenkov light flashes from muons produced in cosmic-ray neutrino reactions in the ocean to detect these neutrinos. This method is an essential component of the proposed project DUMAND to detect cosmic neutrinos,[2] as well as of the neutrino communication scheme discussed in this paper.

As is well known, neutrinos have been extensively used, or their use has been proposed, for many fundamental scientific purposes. Studies of their handedness and of the distinction between ν_μ and ν_e have clarified the weak interaction structure. Their observation deep underground has verified the atmospheric-neutrino production mechanism, and high-energy accelerator-produced neutrino beams have been used to study neutral currents, charmed particles, and heavy leptons. Problems which are still actively being investigated by means of neutrinos are the existence of intermediate bosons, the structure of the sun, and the detection of supernova explosions. Outside of Project DUMAND, mention should also be made of the fundamental experiments now completed,[3] in progress,[4] or in the proposal stage,[5] to investigate the possible existence of neutrino oscillations, e.g. conversions of ν_μ into ν_e and vice versa. This phenomenon is relevant to the purposes of neutrino communication, as will be discussed later.

Typical expected detection rates for natural neutrino sources have been quoted,[6] using 10^{11}-ton detectors, as ~30 counts/yr for neutrinos from radio-galaxies (Centaur A); and, using 10^9-ton detectors, as ~400/yr for intergalactic neutrinos, and ~10^4/yr

*This is an expanded version of the invited talk by Dr. F. J. Kelly at the present symposium, which summarized conclusions of the NRL neutrino communication team.

**Also at Catholic University, Washington, D.C. 20064.

ISSN:0094-243X/79/520113-16$1.50

for atmospheric neutrinos. For a 200-ton detector searching for neutrino oscillations in the Fermilab beam at $\sim 10^3$ km distance, the counting rate would be $\sim$0.3/day.

Besides these scientific uses, can neutrinos be employed for practical purposes? Volkova and Zatsepin[7] have proposed to probe the earth's core using neutrinos (in a fashion similar to Alvarez's "x-raying" the pyramids with cosmic-ray muons[8]), noting that the earth will begin to appear opaque to neutrinos of $\gtrsim$30 TeV energy. Such neutrinos could be produced by 100-200-TeV proton accelerators, which may possibly be available thirty years from now. Of more immediate practical interest is the suggestion of Mikaelyan[9] of a remote reactor-diagnostic technique, based on counting reactor antineutrinos via the reaction $\overline{\nu}_e + p \rightarrow n + e^+$, first detected by Reines and Cowan.[10] This technique could be important for monitoring the leakage of fissioning materials and for controlling the proliferation of nuclear weapons.

In the early 1960s, Schwartz[11] in unpublished work suggested the use of accelerator neutrinos for communications puposes. Arnold seems to have been the first to have made this suggestion in print, at the end of an article[12] chiefly concerned with telecommunications by muon beams.[13] The present authors have carried out quantitative studies of the scientific feasibility of neutrino communication, considering neutrino beams from both the present Fermilab accelerator and from its future upgraded version with 1 TeV proton energy ("Tevatron"). Only brief reports[14] of these studies have been given. The present paper contains a more detailed discussion. One of our principal conclusions is that counting rates of $\sim 10^4$ signal counts/hr, sufficient for a low-data neutrino communications link over distances of 3×10^3 km (10^4 km) could be achieved by using the Tevatron and photomultiplier array of side length $\sim$200 m ($\sim$400 m), which would survey an ocean volume of $\sim 10^7$ tons ($\sim 10^8$ tons).

We do not envision that neutrino communication will ever replace electromagnetic and acoustic methods of long-distance communication, but rather that it will supplement these methods for certain specialized needs. This involves low-data-rate telecommunication with a buried or submerged detector with which it would be difficult to communicate otherwise. Since narrowly collimated neutrino beams would be employed, one would be restricted to point-to-point communications, but these could be made to possess a high degree of privacy and absence of message interception. Two further desirable properties of neutrinos for communication are the impossibility of blocking their propagation (in contrast to electromagnetic waves) and the fact that they would not constitute an environmental hazard.

The organization of the present article is as follows. The second section deals with accelerator neutrino sources, neutrino detectors, and counting rates and background. The third section is devoted to communications considerations. Finally, relevant intensity and event-rate formulas for the ideal case of "perfect focusing" of the accelerator-produced charged mesons that decay into neutrinos are summarized in the Appendix.

HIGH ENERGY NEUTRINO SOURCES AND DETECTORS, COUNTING RATES, AND BACKGROUND

A. Neutrino Sources

Since the interaction cross section of muon neutrinos (the type most copiously produced by high-energy accelerators) in the charged-current reactions

$$\nu_\mu(\overline{\nu}_\mu) + N \rightarrow \mu^-(\mu^+) + \text{hadrons} \tag{1}$$

by which they are mainly detected increases linearly with neutrino energy up to the highest measured energies,[15] use of neutrinos with maximum obtainable energies is preferable for telecommunication purposes. Hence, it is desirable to employ as neutrino sources accelerators such as the existing proton synchrotrons at the Fermi National Accelerator Laboratory (FNAL), Batavia, Illinois, and at CERN, Geneva, Switzerland. Both of these produce proton beams of 400 GeV energy. The energy of the FNAL accelerator is to be raised to 1 TeV by 1980.[16] Designs for a 2-5-TeV accelerator are under way at Serpukhov, USSR, and the possibilities of a very big accelerator (VBA) of 10 TeV have been explored.[17]

Table I shows the relevant parameters of these accelerators, the first row giving the proton energy E_p. The existing accelerators are located in ring-shaped tunnels, with a ring radius of 1 km, and a tunnel diameter of 3 m. At FNAL, the circulating proton beam is extracted once every eight seconds in a 20-μsec pulse of intensity $\sim 2 \times 10^{13}$ protons per pulse, which is directed into a metal target, where it produces π and K mesons, as well as other hadrons. The mesons, focused in a magnetic horn, decay mainly into muons and neutrinos while passing through a 400 m tunnel. The muons are absorbed by a 1-km earth shield and in this way an essentially pure neutrino beam is obtained. Its full opening angle is $\vartheta_\nu \approx 3$ mrad. The energy E_ν of these neutrinos is concentrated in an interval of 5-50 GeV, with a maximum located at ~20 GeV, the total flux being about 10^{10} neutrinos per pulse. Fig. 1 shows schematically how such an accelerator could be used for neutrino telecommunication, with the decay tunnel aimed in the desired direction for point-to-point communications, and Fig. 2 depicts the neutrino beam's traversal of the earth and reception by the detector. Note that with $\vartheta_\nu \approx 3$ mrad, the beam diameter is ≈38 km on the opposite side of the globe.

Table I. Parameters of existing and future high-energy proton accelerators and of their neutrino beams. "Tevatron" designates the energy-doubled FNAL accelerator, and the last column (VBA) refers to a very big accelerator.

	Present accelerators		Future accelerators		
	FNAL	CERN	Tevatron	Serpukhov	VBA
E_p (TeV)	0.4	0.4	1	2-5	$\geqslant 10$
protons/pulse	2×10^{13}	10^{13}	5×10^{13}		$\sim 10^{13}$
E_ν (GeV)	5-50	5-50	10-80		40-120
(E_ν at max. (GeV))	(20)	(20)	(35)		(80?)
ϑ_ν (mrad)	3	3	1		0.2

B. Neutrino Detectors

The neutrino message may be received by a detector sensitive to the reactions (1). As mentioned, massive targets are required for obtaining reasonable counting rates. With the possible exception of scintillation detectors, the best detection scheme for neutrinos in the energy range considered here appears to be the Cowan-Überall method,[1] alluded to above. In this method, one uses a very large (10^6 tons or more) body of water (in the ocean, a deep lake, or a flooded mine) as the target as well as the detector. The positive (negative) muons produced by the relevant reaction (1), which on the average carry off two thirds (one half) of the original neutrino energy, propagate a mean distance of 67 m (50 m) in the water, assuming a mean neutrino energy of 20 GeV. All along their path, they emit a cone of Čerenkov light of forward opening angle 41°, at a rate of ⩾ 200 photons/cm in the blue-green spectral region (Fig. 3). This light in turn propagates over a length of 20 m or more in clear water and can be trapped in a system

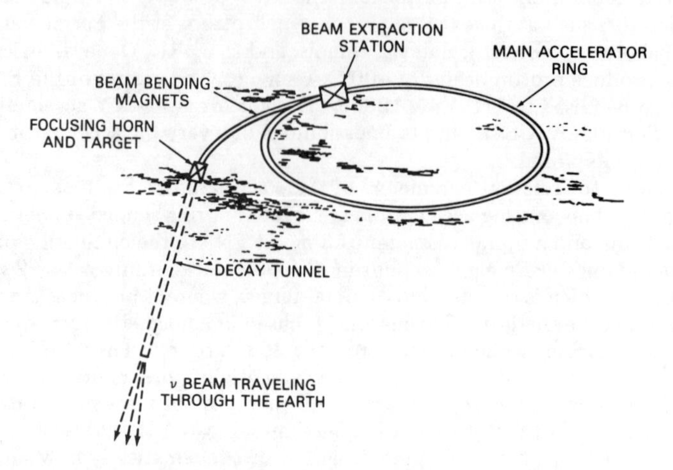

Fig. 1. Scheme for generating a neutrino beam for telecommunication.

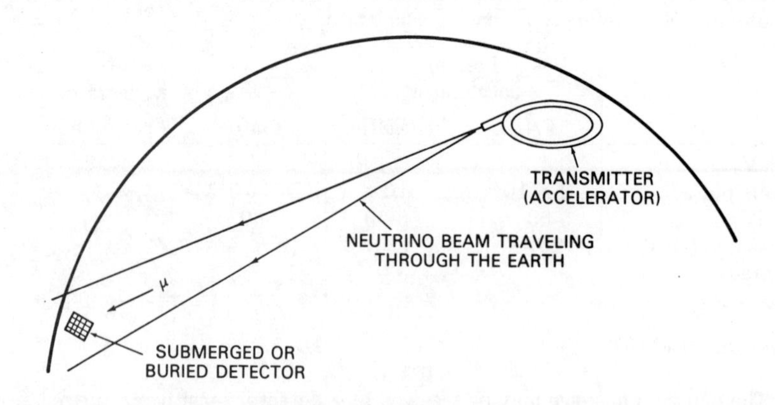

Fig. 2. Neutrino telecommunication with a submerged or buried detector.

of light collectors (e.g., lucite plates or rods) with attached photomultipliers which register the light flash of the muon.

Because of the limitation imposed by the short absorption length of visible light in water, we envisage a large cubic array of detector modules (Fig. 4), spaced 20 m (or at most 40 m) apart. Each module would consist, for example, of a 1-m^2 horizontal lucite plate containing a wave-shifter with one (pressurized) photomultiplier attached to its face or edge. (An analogous detector array has also been proposed for Project DUMAND.) In the first Ref. 14, it was estimated that such a detection scheme would have a muon detection efficiency of close to 100%. The effective target volume of the detector is larger than that of the array, because a muon produced by each of reactions (1) can travel a significant distance in water from its point of origin to where it can be detected by the array.

C. Counting Rates and Background

A program for computing the angular distribution of flux in a neutrino beam and the neutrino event rates induced by the beam at large distances from the accelerator has been developed at NRL. It assumes perfect focusing of the pions and kaons produced by p-nucleus collisions in the accelerator target and is similar to, but simpler than the NUADA program[18] for perfect focusing. The relevant formulas of our program, stated in the Appendix, are based on the Wang spectrum of pion production[19,20] and on a crude modification thereof[21] for kaon production. Here and henceforth in this section, neutrino-event rates (muon-production rates) should be understood as the combined rates for both reactions (1). If the slow variation of the pp absorption cross section with energy is neglected, our formulas yield the expected proportionality of the neutrino event rates per unit volume with E_p^3 when the ratio (decay tube length/E_p) is kept fixed. Roughly speaking, two of the powers of the energy E_p of the accelerator protons arise from the fact that ϑ_ν is inversely proportional to the mean neutrino-beam energy, which in turn is proportional to E_p, and the remaining power of E_p stems from the linear dependence of the total cross section of reactions (1) on the neutrino energy.

In Table II, we show the event rate C predicted by our program for various masses M of water at the indicated distances R from the accelerator source, for various values of the pertinent accelerator parameters, assuming for each E_p considered that the target length equals the absorptive p-nucleus interaction length in the target (maximally efficient target). Here, C = 1/3 × (maximum number of ν-induced events per unit mass per pulse of 5×10^{13} protons, for perfect focusing) × M. The 1/3 factor is intended to simulate "good", but not perfect focusing. Each of the water masses considered is surveyed by a cubic photomultiplier array of side length L_d, centered at the neutrino-beam axis, where the maximum event rate occurs for a given R. Since the neutrino flux is basically uniform throuhout the surveyed water masses at the energies and very small L_d/R values of interest here, the event rates predicted for perfect focusing are close to the respective theoretical maxima for the cases mentioned in Table II. Hence, we believe that the values of C in this table are realistic upper bounds on the event rates in these cases.

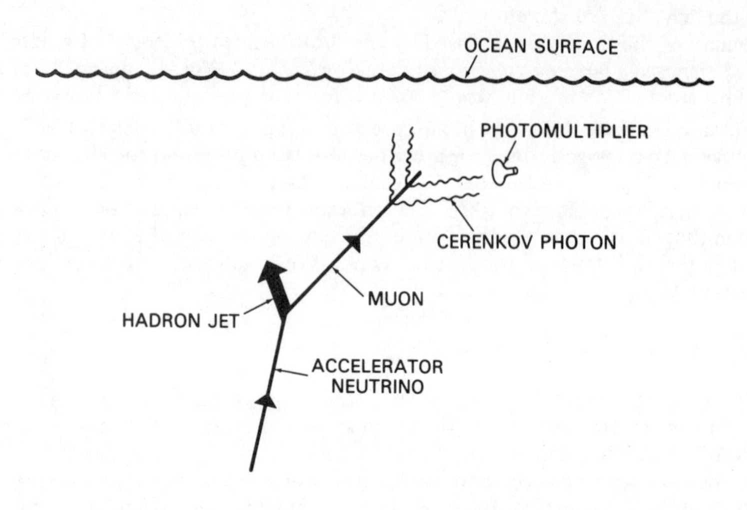

Fig. 3. Čerenkov neutrino detection process.

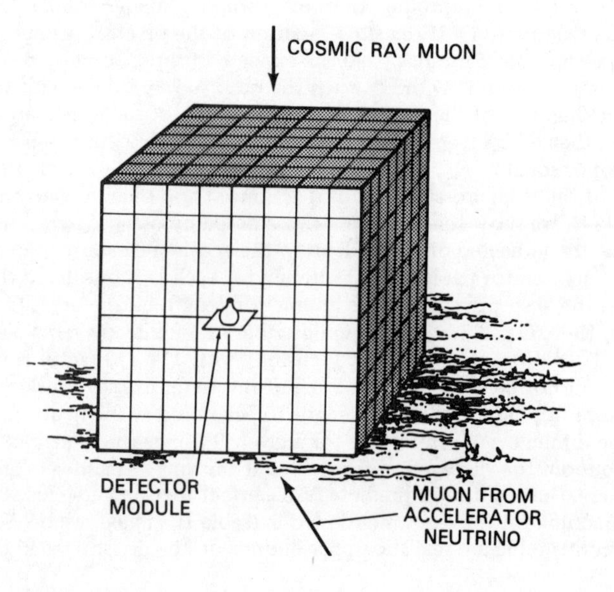

Fig. 4. Čerenkov neutrino detector array. This drawing is intended to be schematic only. An actual array could conceivably consist of detector modules strung along a series of cables anchored to the ocean floor and held vertical by floats on top. Such an arrangement would also provide an open volume of ocean water below the array, to increase the effective volume surveyed for the purpose of detecting upward-traveling neutrinos.

Table II. Calculated neutrino-induced event rates (muon rates) for various distances, accelerator parameters, and detector masses. C = 1/3 × (maximum number of ν-induced events per unit mass per pulse of 5×10^{13} protons) × M at a distance R from the accelerator, assuming perfect focusing and target length = p-nucleus absorptive interaction length; M = water mass surveyed by a cubic photomultiplier array of side length L_d; N = number of detector modules of the array. Distance between modules = 20 m.

Accelerator energy and decay-tube length	M (tons)	L_d (m)	N	$R = 3.2 \times 10^3$ km C (events/pulse)	$R = 10^4$ km C (events/pulse)
0.4 TeV, 0.4 km[a]	10^7	200	1,000	0.85	0.085
	10^8	450	11,000	8.5	0.85
1.0 TeV, 0.4 km[a]	10^6	70	40	0.53	0.053
	10^7	180	730	5.3	0.53
	10^8	430	10,000	53	5.3
1.0 TeV, 1.0 km[a]	10^6	70	40	1.3	0.13
	10^7	180	730	13	1.3
	10^8	430	10,000	130	13
2.5 TeV, 2.5 km[b]	10^6	50	20	20	2.0
	10^7	150	420	200	20

[a] For this case, C was computed using Eq. (10) and its analogues for the π^- and $K^\pm$ contributions. The needed values of the pp absorptive cross section were obtained from U. Amaldi et a., Phys. Lett. 44B, 112 (1973).

[b] The values of C shown in this case are simply $(2.5)^3$ times larger than the corresponding values for E_p = 1 TeV and a 1-km decay tube.

Assuming the present Fermilab repetition rate of 450 proton pulses/hr, and that 5×10^{13} protons/pulse are delivered to the accelerator target, the event rates in Table II entail the following hourly rates. For the present Fermilab proton energy (0.4 TeV) and decay-tube length (0.4 km), 3.8×10^3 muons/hr would be produced in a water detector with $M = 10^8$ tons at $R = 3.2 \times 10^3$ km. This rate would suffice for neutrino communication at this distance. These muons could be detected by a photomultiplier array with $L_d \sim 450$ m. At $R = 3.2 \times 10^3$ km (10^4 km), we predict that a 1-TeV accelerator with a 1-km decay tube would yield 5.8×10^3 muons/hr in a 10^7 ton (10^8 ton) water volume, surveyable by an array with $L_d \sim 180$ m ($\sim$430 m); and that a 2.5-TeV accelerator with a 2.5 km decay tube would produce 9.0×10^3 muons/hr in a 10^6 ton (10^7 ton) water volume, surveyable by an array with $L_d \sim 50$ m ($\sim$150 m). These figures reflect, of course, the $\sim E_p^3$ behavior of the event rates/unit volume and the proportionality with E_p of the mean muon range in water (conservatively, $\gtrsim$300 m for a 2.5-TeV accelerator with a 2.5-km decay tunnel).

Since the muons produced in the reactions (1) emerge at rather small angles with the beam direction at the E_p values of interest here, and in view of the above event-rate and muon-range dependence on E_p, we arrive at the following result for such energies. The event rate in the volume of water surveyed by a planar photomultiplier (or scintillator) array in the ocean, say perpendicular to the beam (to maximize the volume), would be approximately proportional to E_p^4, for given values of R, the (decay tube length/E_p) ratio, the number of protons/pulse, and the accelerator repetition rate. This elementary conclusion could be very relevant for global neutrino communication by means of a proton accelerator in the multi-TeV range.[22]

Background to the neutrino-induced signals is provided mainly by sunlight, Čerenkov light from cosmic-ray muons, and bioluminescence. Flashes of the latter origin typically last for milliseconds, and can thus be discriminated against the nanosecond Čerenkov flashes from the muons produced in reactions (1). On the basis of known data,[23] we estimate that for 10^4 signal counts/hr immersion depths of 300-400 m and 600-700 m would provide adequate shielding against cosmic-ray muons and sunlight in the ocean, respectively (of course, sunlight could be eliminated by enclosures). With higher counting rates, the immersion depths could be correspondingly reduced.

COMMUNICATIONS CONSIDERATIONS

The laws of physics and the ingenuity of accelerator designers have provided a means for sending neutrinos through the earth to a detector which can sense the presence of muons produced by neutrinos and register a signal. It remains to select a method of modulation of the neutrino beams and a coding procedure which will provide satisfactory transfer of messages with a low error rate in the presence of the natural noise backgrounds. Just as a multitude of modulation and coding schemes have been used for radar and optical (laser) systems, a large variety of modulation and coding schemes are possible for neutrino beams. Ordinarily, selection of modulation techniques is related to the technological ease and economical benefits of a system, in addition to any crypto-security and anti-jam protection that need be built in. Moreover, since a message usually is passed through several propagation media and several message centers, the detailed character, word, and page structure of a message-handling system must be considered in the design of an ultimate coding scheme. However, it is still instructive to discuss a basic method of using a neutrino beam to convey information and to show how the rate of information transfer — the channel capacity of the communications link — can be evaluated.

The communications system concepts that are relevant to pulsed-neutrino communications are very similar to those governing pulsed-laser optical communication which have been treated previously in the communications literature. The neutrino-communication technique is unique in the penetrating ability of the beam and the long buildup time between short pulses of neutrinos. The pulse-position-modulation technique for transferring information consists of placing a pulsed signal into one of a large possible sequence of time intervals. The application of pulse-position modulation to laser-beam communication has been considered by Gagliardi et al.,[24] according to whom the theoretical limit for the data rate H in bits per second, assuming perfect coding for an M-ary symmetric channel, is

$$H = \frac{\log_2 M + P_E \log_2 [P_E/(M-1)] + (1-P_E)\log_2 (1-P_E)}{M\Delta T} . \quad (2)$$

Here, M is the number of time intervals per pulse; P_E is the probability that an error will occur in the reception of a pulse, i.e., the probability that a receiver will make a mistake in the estimation of the time interval in which the pulse was emitted; and ΔT is the pulse duration.

The following example shows how a neutrino-communication system might employ this method. We assume that a transmitting accelerator and a receiving array have been constructed, such that 10^4 neutrino events per hour (22 events per 8-sec cycle) may be reliably observed in the detector. We also assume that the synchrotron transmitter of 1-km radius requires 7 sec to bring a proton beam to its maximum energy.[25] It will then take $\sim 2 \times 10^{-5}$ sec for the entire circulating beam to emerge from the accelerator after an internal magnetic-field "kicker" is triggered. It is not difficult to ensure that the transmitter and receiver have synchronized clocks and that the receiver takes into account the propagation delay between the triggering of a kicker and the receipt of a neutrino pulse. (At the beginning of each communication period, several synchronizing neutrino pulses can be sent to provide this clock alignment automatically.) After each subsequent 7-sec acceleration period (see Fig. 5), the transmitter is prepared to send its message. Each eighth second may be divided into 32768 ($=2^{15}$) equal intervals of time, each of duration 3.05×10^{-5} sec. A unique 15-bit binary message may be associated with each time interval as shown in Fig. 5. The time at which the kicker is triggered is selected according to the binary message to be sent. After the triggering event, the neutrinos are generated and travel through the earth with the speed of light to the receiving array. There, the neutrino pulse causes about 22 events to occur in the array during the 2×10^{-5} sec interval corresponding to the proton beam-spill time. Clocks and counter electronics are arranged to decode the observed time of occurrence of these events and hence to reconstruct the 15-bit binary message. Using Eq. (2) corrected for a 7-sec dead time, and assuming $P_E \approx 0$, we find that the system would then operate at a communications rate of ~ 1.9 bits/sec or 15 bits per 8 sec. An electronic clock having an accuracy of one part in 10^8 would maintain adequate synchronism for a communication period of 3000 sec.

It should not prove difficult to construct and deploy a receiving array in which false neutrino pulse events are unlikely to arise spontaneously from the background noise. Using a simple detection criterion which interprets the presence of eight or more events within the detector array during a 3×10^{-5}-sec interval as a valid neutrino signal, the probability of message error will be less than 10^{-3} if the average single neutrino-like (false) event rate is less than 0.4 events per 3×10^{-5}-sec interval at 1.3×10^4 events/sec. For the above mentioned example of a 10^8-ton water detector which detects an average of 22 neutrino events for each pulse of the accelerator, a message-error rate of $P_E = 10^{-3}$ could be achieved by immersion to about 300 m, assuming that cosmic-ray muons furnish the only significant background, i.e., assuming that sunlight has been excluded by suitably covering the detector. For the same array conditions given above, an error rate of $P_E = 10^{-5}$ for each 15-bit message could be achieved by reducing the threshold for pulse detection to 5 neutrino events and by increasing the depth of the array to about 1000 m, to reduce the accidental cosmic-ray background sufficiently.

EFFECT OF NEUTRINO OSCILLATIONS ON NEUTRINO COMMUNICATIONS

Of the many postulated explanations for the unexpectedly low rate of the measured neutrino signal from the sun,[26] one of the most natural is the hypothesis of neutrino oscillations, a theoretically possible phenomenon first discussed by Pontecorvo in 1957.[27] Various aspects and ramifications of this hypothesis, or of its modifications and consequences, are discussed in several papers in these proceedings.[28] If muon neutrinos oscillate into other neutrino types, such as electron neutrinos, this could entail major consequences for the proposed neutrino-communication scheme. Thus, a few words relative to such consequences are in order here.

A necessary condition for neutrino oscillations to take place in vacuum is that the mass of at least one of the participating neutrino types is nonvanishing. Pontecorvo predicts an appropriate "oscillation length" L, which is proportional to the neutrino momentum. Neutrino communications would be carried out with accelerator-neutrino beams, and hence with almost pure muon neutrino beams whose energy spectrum would typically be rather wide. It has been predicted for such a spectrum that a reduction of

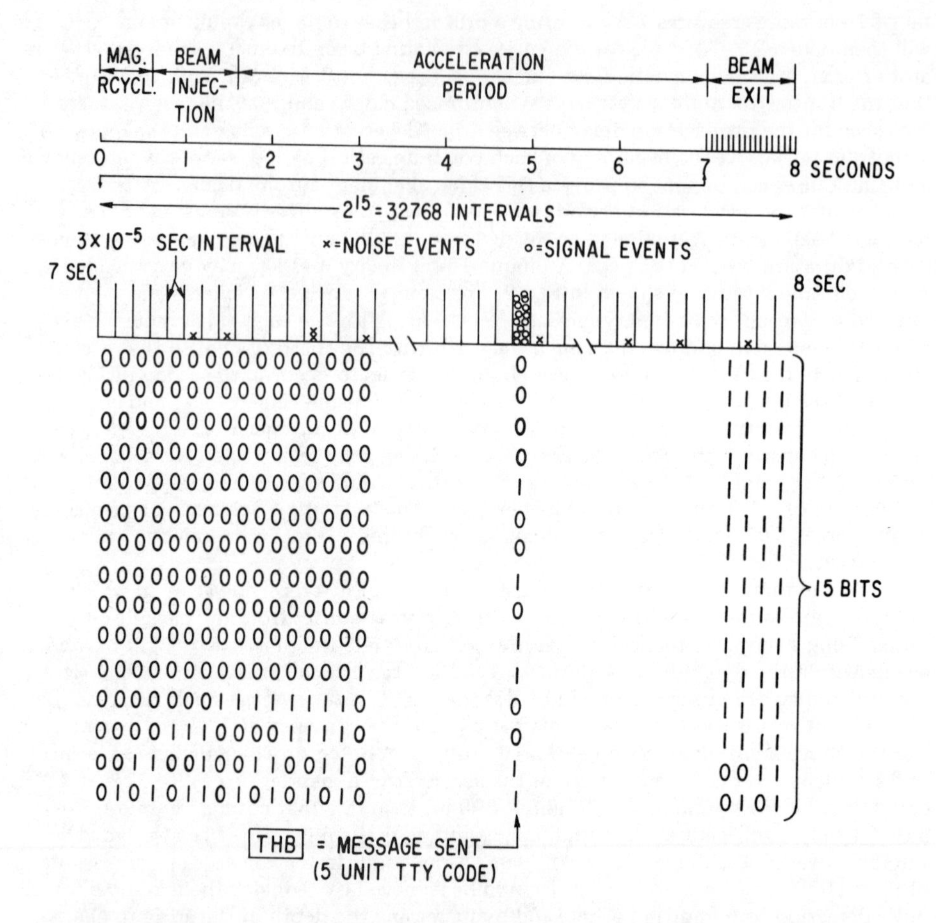

Fig. 5. Pulse-position modulation. The upper portion of the figure shows the synchrotron repetition period for the systems described in the text. First, there is a 0.6 second initial period in which the ring magnet is returned from its highest field to its initial field condition. Then, the protons are injected into the synchrotron over the period of a second. The protons are accelerated to their maximum energy during the following 5.4 seconds. During some short interval between the seventh and eighth seconds the ejection of the proton beam from the synchrotron by the use of a "kicker" magnet causes neutrino events in the corresponding time interval at the detector, thereby transmitting the three letters THB in international telegraphic code.

the muon neutrino content of the accelerator beam by a factor as small as $\geqslant 1/n$, relative to its value in the absence of the oscillations, will occur at distances long compared to the value of L averaged over the energy spectrum, if there exist n types of neutrinos (including muon neutrinos) which oscillate into one another (see the first Ref. 5). The discovery of a new type (tau) of lepton[29] suggests the existence of an associated neutrino type, and hence that $n \geqslant 3$, and astrophysical evidence[30] suggests an upper limit of seven for the total number of neutrino types.

While this means that the muon-production rate in a distant volume of the ocean illuminated by an accelerator-neutrino beam and surveyed by a lattice of Čerenkov detectors could simply get reduced by an overall factor $\geqslant 1/n$ as compared to the predictions of calculations (such as the present ones) which do not take neutrino oscillations into account, much more complicated effects (and, in certain cases, more serious reductions in the muon-production rate) may take place if neutrino oscillations are affected by the presence of matter, as discussed by Wolfenstein.[31] This effect of matter comes about through an index of refraction for neutrinos in matter, which is caused by the elastic forward-scattering of neutrinos from atomic electrons via the neutral weak current. Under reasonable assumptions, the effect in question leads to a phase change of 2π of the neutrinos over distances $\sim 10^4$ km, i.e., of the order of the earth's diameter.

This effect of matter may change the Pontecorvo-type (vacuum) neutrino oscillations depending on the length and density of matter traversed. However, the presence of matter may induce oscillations of neutrinos even if the latter are massless, provided the neutral current connects different neutrino types. With reasonable assumptions, Wolfenstein has predicted quite dramatic results for the probability of a ν_μ remaining a ν_μ when traversing the earth, as shown in the figure of his paper in these proceedings.[31] A partial or even essentially complete loss of ν_μ intensity may take place in an accelerator-neutrino beam transmitted through the earth in certain directions, with no significant loss occurring for others. Hence, neutrino communication would encounter certain "blind spots" if this property of the neutral current and the other conditions stated in Refs. 31 obtain. Since the matter-induced neutrino oscillations in question are independent of neutrino momentum, these directional effects would not be washed out by averaging over the accelerator-neutrino energy spectrum.

Therefore, the experimental investigation of the existence of both the vacuum and matter-induced neutrino oscillations is highly relevant for neutrino telecommunication purposes. The experiment proposed by Lande et al.,[32] namely, to detect atmospheric muon neutrinos which have traveled up to a full diameter through the earth, would show whether neutrino oscillations in matter exist at a significant level over global distances and how they differ from vacuum oscillations.

CONCLUSIONS

Telecommunication over global distances by means of neutrino beams is proposed as a supplement to the conventional electromagnetic-wave communication methods. Assuming the use of suitable underwater Čerenkov detectors (and barring serious complications due to neutrino oscillations), neutrino telecommunication is shown to approach feasibility if presently existing high-energy proton accelerators are employed as neutrino sources, and to be definitely feasible with the advent of higher-energy accelerators which are already in the design stage. Special advantages of this type of communications, as compared to other types, are that they could be made essentially safe from blockage and disruption, as well as to furnish a high degree of privacy. The method may prove useful as a low-data link to buried or submerged receivers with which communications might otherwise be difficult.

The authors wish to thank the following NRL scientists for helpful discussions: R. H. Bassel, H. Beck, M. Hass, J. A. Murray, M. Rosen, N. Seeman, M. M. Shapiro, and W. W. Zachary.

APPENDIX

NEUTRINO INTENSITIES AND INDUCED EVENT RATES AT LARGE DISTANCES

The formulas for the neutrino intensities and event rates given in this appendix apply to the case of "perfect focusing" of the pions and kaons generated in high-energy p-nucleus collisions in the accelerator target. As mentioned Sec. 2, they are based on the Wang spectrum[19,20] of pion production by such collisions[33] and on the usual modification[21] of this spectrum to obtain kaon production by the incident proton beam. We feel that our formulas, which are specifically intended for predicting neutrino intensities and induced event rates at large distances from the source, are simpler and more convenient for "everyday" use than the NUADA perfect focusing program.[18]

We first consider ν_μ production by $\pi^+ - \mu^+$ decay of the π^+ mesons generated in p-nucleus collisions in the accelerator target. The target is supposed to be cylindrical, with the incident proton beam directed along an interior line parallel to the cylindrical axis, and to be composed of a single, sufficiently light nuclear species, such as Be or Al. (The same formulas apply to $\bar{\nu}_\mu$ production from the decay of the π^- from these collisions, with obvious interpretations of the appropriate constants). Assume that the π^+ mesons are focused immediately after emerging from the target, so that their momentum vectors point in the direction from 0 to 0′ (Fig. 6), 0 being the accelerator site. Let P be a point of the plane through 0′ perpendicular to 00′ (Fig. 6) and suppose that the distance R from 0 to 0′ is much larger than the length r of the accelerator's pion and kaon decay tunnel. For incident proton energies of hundreds of GeVs, the customary assumption of highly relativistic pions and the small-angle approximation (sine ≈ argument) made throughout the appendix are justified. In addition, we make the usual approximation[17] of taking the absorptive interaction lengths of the pions and kaons in the target as equal to that of the primary protons.

Under these circumstances, the number of ν_μs per accelerator pulse per unit area at P is

$$N_\nu(\mu) = kI_1(\mu)(E_p^2/R^2), \qquad (3)$$

where μ is the cosine of the angle between the neutrino path OP and the neutrino beam axis 00′, E_p is the incident proton energy in GeV, and k and $I_1(\mu)$ will now be defined. The quantity k is in $(\mathrm{GeV})^{-2}$ and is defined by

$$k = 2\ell e^{-\ell} N_p (A'/m^2c^4D^2), \qquad (4)$$

where N_p is the number of protons entering the target per pulse, ℓ the target length in units of the p-nucleus absorptive-interaction length in the target at energy E_p, m the pion mass in GeV/c^2, c the speed of light, $A' = A/\sigma_a$, where σ_a is the pp absorptive cross section at E_p in mb, and A and D are Wang's parameters[19] for π^+ production. Recall that A′ is in units of $(\mathrm{GeV/c})^2$, that Wang's parameters B and C for π^+ production mentioned below are dimensionless, and that D is in $(\mathrm{GeV/c})^{-1}$. We define the dimensionless quantities

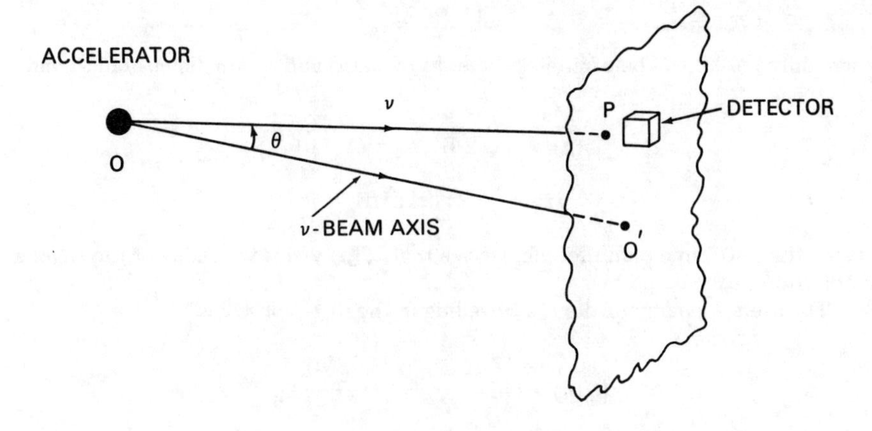

Fig. 6. Neutrino beam and detector geometries.

$$I_j(\mu) = \left| \int_0^1 \frac{z^j(1-z)e^{-Bz^C}(1-e^{-(z/z_0)})}{[1+a^2(\mu)z^2]^{j+1}} dz, \right. \tag{5}$$

where only j = 1,2 will be of interest here. In (5), $a(\mu)$ and z_0 are dimensionless and given by

$$a(\mu) = (\sqrt{2}\, E_p/mc^2)(1-\mu)^{\frac{1}{2}}, \tag{6}$$

$$z_0 = (r/c\tau)(mc^2/E_p), \tag{7}$$

τ being the rest-frame pion lifetime. Notice that $I_1(\mu)$ yields the shape of the angular distribution of the ν_μ beam.

The average energy of the ν_μs travelling in the direction OP is

$$E_\nu(\mu) = \left(\frac{m^2 - m'^2}{m^2}\right) \frac{I_2(\mu)}{I_1(\mu)} E_p, \tag{8}$$

m' being the muon mass in GeV/c^2.

The number of reactions

$$\nu_\mu + N \rightarrow \mu^- + \text{hadrons} \tag{9}$$

at P per accelerator pulse per unit volume due to the ν_μs from the $\pi^+ - \mu^+$ decay tube is

$$C_\nu(\mu) = k' I_2(\mu)(E_p^3/R^2), \tag{10}$$

with

$$k' = kN_d\alpha_\nu\left(\frac{m^2 - m'^2}{m^2}\right). \tag{11}$$

Here, N_d is the nucleon number density at P and $\alpha_\nu = 0.61 \times 10^{-38}$ cm^2/GeV is the proportionality constant in the formula $\sigma_\nu = \alpha_\nu E_\nu$ for the total cross section of reaction (9) in the GeV range.[15]

We use the crude rule[21] that the flux $d^2N/dpd\Omega$ of K^+s(K^-s) generated per absorptive p-nucleus interaction is equal to 0.15 (0.05) times the corresponding flux $d^2N/dpd\Omega$ of π^+s (π^-s) generated in such an interaction at a given proton energy E_p. Under this assumption, Eqs. (3), (8), and (10) are easily modified to apply to the kaon case. The kaon contribution to the count rates shown in Table II is less than 20% of the pion contribution. This is to be expected, since the mean angular divergence of the neutrinos from K-μ decay is much larger than that of those from π-μ decay and since the only neutrinos from K-μ decay contributing to the rates in question are those which have emerged at angles of $\leqslant 10^{-4}$ radians.

REFERENCES

1. H. Überall and C. L. Cowan, in Informal Conference on Neutrino Physics, Geneva, Jan. 20-22 (1965), CERN Report No. CERN 65-32 (unpublished); C. L. Cowan, H. Überall, and C. P. Wang, Nuovo Cimento 44, 526 (1966); G. K. Riel, C. L. Cowan et al., Bull. Am. Phys. Soc. 12, 1075 (1967).
2. See the articles in Chap. 2 of these proceedings for a discussion of various aspects of Project DUMAND and appropriate references.
3. See, e.g., H. W. Sobel and F. Reines, University of California, Irvine, preprint (unpublished).
4. See, e.g., L. R. Sulak et al., Neutrino '77, proceedings of the International Conference on Neutrino Physics and Neutrino Astrophysics, Baksan Valley, U.S.S.R., 1977, edited by M. A. Markov (Nauka, Moscow, 1978), Vol. 2, p. 280.
5. See, e.g., A. K. Mann and H. Primakoff, Phys. Rev. D 15, 655 (1977), and K. Lande et al., Neutrino '77, Vol. 1, p. 170.
6. M. M. Shapiro, unpublished lectures.
7. L. V. Volkova and G. T. Zatsepin, Bull. Acad. Sci. USSR, Phys. Ser. 38, 151 (1974).
8. L. W. Alvarez et al., Science 167, 832 (1970).
9. L. A. Mikaelyan, Neutrino '77, Vol. 2, p. 383.
10. F. Reines and C. L. Cowan, Jr., Phys. Rev. 113, 273 (1959).
11. We wish to thank Dr. A. Roberts and Prof. H. Faissner for informing us of this proposal of Prof. M. Schwartz.
12. R. C. Arnold, Science 177, 163 (1972).
13. In addition to these modern workers, Roman mythology recounts the tale of a precursor of future transterrestrial communication in the well known narrative (Ovid, Metamorphoses, translated by F. J. Miller (Harvard University Press, Cambridge, Massachussetts, 1958), Book XI, p. 133), of King Midas' barber who whispered a secret into a hole in the ground. Reeds sprang up from the ground and whispered the secret to the wind, which spread it far and wide.
14. A. W. Sáenz et al., Science 198, 295 (1977); Bull. Am. Phys. Soc. 23, 545 (1978); Neutrinos-78, proceedings of the International Conference on Neutrino Physics and Astrophysics, West Lafayette, Indiana, 1978, edited by E. C. Fowler (Purdue University, 1978), p. C168.
15. B. C. Barish et al., Phys. Rev. Lett. 39, 1595 (1977).
16. R. R. Wilson, Physics Today 30, 23 (1977); 1976 Summer Study, Utilization of the Energy Doubler/Saver, edited by J. Lach (Fermilab, Batavia, Illinois), Vols. 1 and 2.
17. See, e.g., J. B. Adams, in CERN Annual Report, 1976; L. Camilleri, CERN Report No. CERN 76-12, June 1976 (unpublished).
18. D. C. Carey, Fermilab Report No. FN-247, October 10, 1972 (unpublished).
19. C. L. Wang, Phys, Rev. D 7, 2609 (1973).
20. C. L. Wang, Phys. Rev. D 10, 3876 (1974).
21. C. Baltay, in Weak Interactions, edited by V. W. Hughes and C. S. Wu (Academic, New York, 1975), Vol. II (Muon Physics), p. 304, Eq. (54).
22. This conclusion, well known to the NRL neutrino communication group, has been independently stated in a report by T. Bowen et al., September 1978 (unpublished), written by several participants of the DUMAND 1978 Summer Workshop.
23. S. Higashi et al., Nuovo Cimento A43, 334 (1966); R. H. Oster and G. L. Clarke, J. Opt. Soc. Am. 25, 84 (1935).
24. Robert M. Gagliardi and S. Karp, IEEE Trans. on Communications Technology COM-17, 208 (1969). See also W. N. Peters and G. S. Entwistle, in Proceedings of the Tenth National Communication Symposium (Utica, New York), p. 94; I. Bar

David, IEEE Trans. on Information Theory IT-15, 31 (1969); B. Reiffen and H. Sherman, Proc. IEEE 51, 1316 (1963); K. Abend, IEEE Trans. Information Theory (Correspondence) IT-12, 64 (1966); T. F. Curran and M. Ross, Proc. IEEE (Correspondence) 53, 1770 (1965).
25. J. R. Sanford, Annu. Rev. Nucl. Sci. 26, 151 (1976).
26. R. Davis, Jr., J. C. Evans, and B. T. Cleveland, these proceedings.
27. B. Pontecorvo, Zh. Eksp. Teor. Fiz. 33, 549 (1957); 34, 247 (1958).
28. See the articles by K. Lande et al., A. K. Mann, L. Wolfenstein, and R. Ehrlich in these proceedings.
29. G. J. Feldman and M. L. Perl, Phys. Rep. 33C, 285 (1977).
30. G. Steigman, D. N. Schramm, and J. E. Gunn, Phys. Lett. 66B, 202 (1977).
31. L. Wolfenstein, Phys. Rev. D 17, 2369 (1978) and article in Chap. 3 of these proceedings.
32. See, e.g., K. Lande et al., these proceedings.
33. In particular, formulas (3), (8), and (10) of the Appendix have been derived under the assumption that, for sufficiently light nuclear targets, the flux $d^2N/dpd\Omega$ of $\pi^{\pm}$ produced per absorptive p-nucleus interaction is approximately the same as the corresponding flux due to pp collisions at the same E_p value. See, e.g., Ref. 20 and the relevant references therein.

TIME DEPENDENCE OF THE BROOKHAVEN SOLAR-NEUTRINO COUNTING RATE AND THE NEUTRINO-OSCILLATION HYPOTHESIS

Robert Ehrlich
Physics Department, George Mason University, Fairfax, Virginia 22030

ABSTRACT

Pomeranchuk has suggested that one might directly observe neutrino oscillations in a solar-neutrino experiment if the oscillation wavelength were comparable to the annual earth-sun distance variation from perihelion to aphelion, 5×10^6 km. We find that data from the Brookhaven solar-neutrino experiment can be interpreted to marginally favor the neutrino-oscillation hypothesis at the 2-standard-deviation level. If the effect is not a statistical fluctuation, the estimated value for $\Delta m^2 = m_{\nu_1}{}^2 - m_{\nu_2}{}^2 \sim 4 \times 10^{-10}$ eV^2 is such that terrestrial tests would seem to be unfeasible.

INTRODUCTION

During the last decade, Davis et al. have been performing a very careful search deep underground for neutrinos from the sun.[1] With the gradual accumulation of data, their results have become more precise and disagreement with the prediction of standard solar models has become more acute, thereby spawning a large number of suggested resolutions of the problem of the "missing" solar neutrinos.[1,2] A recently reported value[3] from Davis' experiment is 1.6 ± 0.4 SNU (1 SNU = 10^{-36} captures per atom per second). This value, which results after subtracting a cosmic-ray background of 0.4 ± 0.2 SNU, differs from a recent standard-solar-model value,[4] 4.7 SNU, by a factor of three. One possible resolution of the missing-solar-neutrino puzzle is based on Gribov's and Pontecorvo's suggestion[5] of neutrino oscillations. Neutrino oscillations arise if at least one neutrino type has nonzero mass, and if muon and electron number conservation is not absolute, permitting the transformations $\nu_e \rightleftharpoons \nu_\mu$. In this case, Gribov and Pontecorvo suggested that some fraction of the solar electron neutrinos would transform into muon neutrinos before reaching earth and go undetected in Davis' experiment. Pomeranchuk[5] has noted that, because of an annual earth-sun distance variation of 3%, it might be possible to directly observe such neutrino oscillations if their wavelength is comparable to 5×10^6 km, the range of the earth-sun distance variation from aphelion to perihelion. The purpose of this paper is to note that recently reported data from Davis' experiment marginally favor the neutrino-oscillation hypothesis, according to this author's interpretation of the data.

INTERPRETATION OF DATA FROM DAVIS' EXPERIMENT

The results of runs 18 - 47 from Davis' ^{37}Cl solar neutrino experiment are shown[6] in Fig. 1. Davis has already suggested[1] that it is conceivable there are time variations present in the data, but that most theorists agree that it is extremely unlikely that the solar neutrino flux would vary. However, as noted above, an annual variation could arise from neutrino oscillations rather than a time-varying flux at the source, and hence such a variation is intrinsically more plausible than any other. The ability to make a statistically meaningful statement about the possible presence of time variations in Davis' data is significantly greater if one combines runs from different years on a single annual

ISSN:0094-243X/79/520129-10$1.50

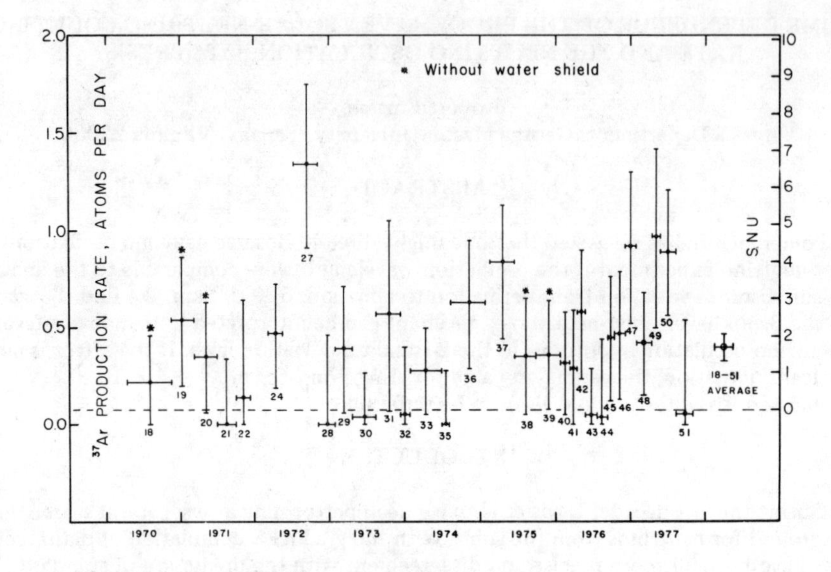

Fig. 1. Solar-neutrino counting rate (in ^{37}Ar atoms per day and SNU) vs. time. Graph is taken from Davis' BNL report (Ref. 3).

time scale. In Fig. 2, we have plotted Davis' data on a single annual scale starting at January 4, the date of perihelion. Plotting the data in this way makes it somewhat easier to visually spot any systematic annual variation. There is, of course, one obvious source of an annual variation, namely, the change in solid angle subtended at the detector during the course of the year. However, this effect is small and the appropriate correction factor, ranging from +3% to −3%, has been applied to the data of Fig. 2 and Fig. 5. If there are annual variations in the corrected data arising from neutrino oscillations as a function of earth-sun distance, then the variations must be symmetrical about January 4 and July 4, the dates of perihelion and aphelion. The data of Fig. 2 do, in fact, appear to be reasonably consistent with symmetry about the date of aphelion. For example, a fit to the function $N(t) = \beta_1 + \beta_2 \cos(t-t_0)$ yields a best value for $t_0 = 13 \pm 46$ days, in good agreement with the expected value $t_0 = 0$ required by symmetry, where t is measured from the perihelion date, $t_0 = 0$.

In testing the hypothesis that there exists an annual variation in counting rate, it is of some significance that seven out of eight runs occurring in the middle of the year fall below the average value. For example, if we compute the average counting rates in the first, second, and third four months of the year (measured from January 4), we obtain:

$$N_1 = 1.9 \pm 0.5 \text{ SNU},$$
$$N_2 = 0.7 \pm 0.5 \text{ SNU},$$
$$N_3 = 1.7 \pm 0.6 \text{ SNU}.$$

This yields for the quantity $\Delta N = \frac{1}{2}(N_1 + N_3) - N_2$ a value $\Delta N = 1.1 \pm 0.6$ SNU that is significantly different from zero, the expected value for a non-varying counting rate.

The evidence against a constant counting rate becomes even more compelling when we examine the statistical treatment of runs with low-counting rates in Davis' experiment. To quote Ref. 3:

> "The counts occurring in the energy (FWHM) - rise time (90%) window corresponding to the Auger electrons from ^{37}Ar decay were resolved by the statistical treatment into a decaying component with a 35 day half-life and a nondecaying background. The number of counts observed is small; therefore the statistical treatment gives the probability distribution function for the number of ^{37}Ar atoms produced in each experimental run. From this distribution function is obtained the most likely value and the 34% confidence ranges above and below this most likely value. In the cases where the most likely value is low and the area under the probability distribution function below the most likely value is less than 34 percent, the upper bound given includes 68 percent of the area under the probability distribution function." (emphasis added)

Thus, the definition Davis uses for the error bars for the low-data points means that if all data points lie off a fitted curve by the same multiple of their error bars, then the low-data points would decrease the probability of a fit by a significantly greater amount than the others. An alternative approach would be to treat high and low-data points in the same manner, showing the 34% confidence ranges above and below the best value of each run. Using this definition, there would be roughly the same contribution to a χ^2 or likelihood function when any given data point lies off a fitted curve by a given multiple of its error bar. It is quite clear that such a reduction in the error bars of the low-data

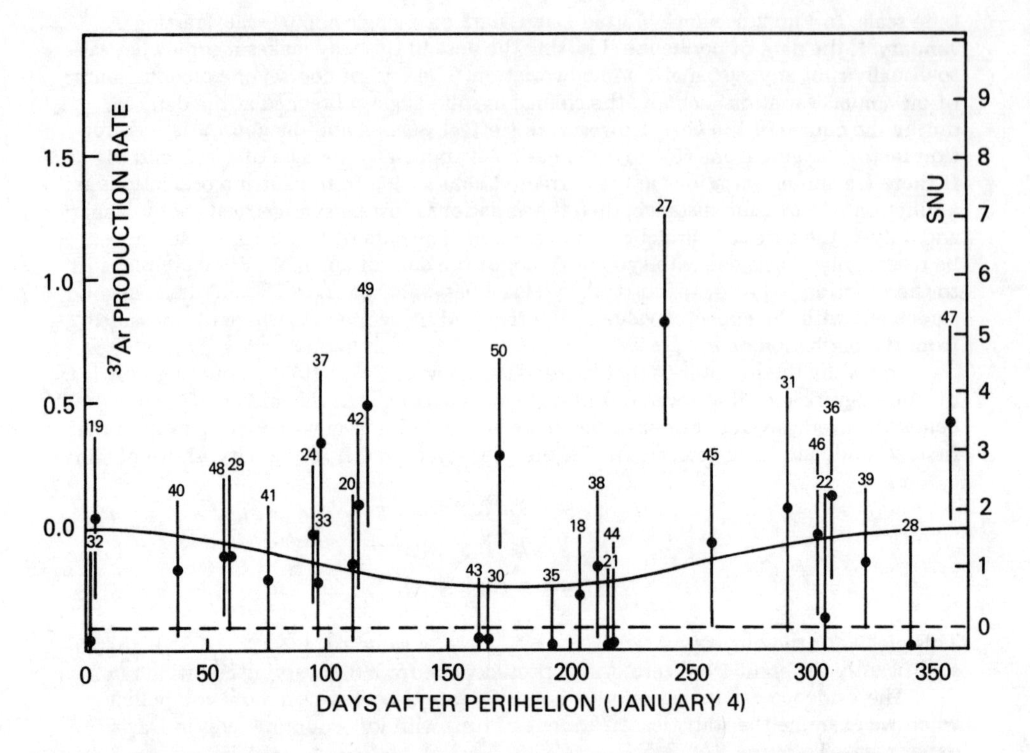

Fig. 2. Davis' data, corrected for varying solid angle subtended by the detector during the course of a year and plotted on a single annual scale commencing at the date of perihelion (January 4). The horizontal bars of figure I, showing the length of time each run lasts, have been omitted for clarity.

points, most of which lie in the middle of the year, would have a significant effect on the goodness of fit for the constant-counting-rate hypothesis.

There is, moreover, a second reason why it may be desirable to treat high and low-data points in identical fashion, and thereby allow both data points and error bars to go below zero. Suppose that the true neutrino-induced ^{37}Ar-production rate were exactly zero. In this case, because of the way the ^{37}Ar rate is extracted, statistical fluctuations in the background may well yield a best value on many runs that is positive. If negative (unphysical) values are not allowed, then a systematic error is introduced when one computes the time-averaged counting rate over all runs.[7] In fact, as the number of runs becomes arbitrarily large, the error in the average value will approach zero, but the average rate will be positive and, therefore, statistically different from zero.

IMPLICATIONS FOR NEUTRINO-OSCILLATION HYPOTHESIS

We shall now consider the implications of an observed varying counting rate for the neutrino-oscillation hypothesis. Further runs of Davis' ^{37}Cl experiment should be able to settle the question of whether the effect discussed here is simply a statistical fluctuation.

Suppose that neutrinos of two given types only oscillate into one another, and that they do this in a maximal way. If only one of the two types is produced initially, the neutrino flux at a distance R from the production site is given by[8]

$$\phi(R) = \tfrac{1}{2}\,\phi(0) \cdot (1 + \cos kR), \tag{1}$$

where $\dfrac{2\pi}{k} = \dfrac{2.5\,E_\nu}{(m_{\nu_1}{}^2 - m_{\nu_2}{}^2)}$ km. Here, $\phi(0)$ is the neutrino flux at a distance R in the absence of neutrino oscillations, E_ν is the neutrino energy in GeV, R is the source-detector distance in km, and m_{ν_1}, m_{ν_2} are the masses in eV of the eigenstates of the neutrino mass matrix defined by Gribov and Pontecorvo.[5] Bahcall[8] has noted that the effect of averaging equation (1) over neutrino energy, E_ν, will smear out any neutrino oscillations for large values of kR and simply result in an average counting rate which is one-half the expected rate in the absence of oscillations. This follows from the fact that the fractional range in energies $\Delta E_\nu/E_\nu$ for solar neutrinos having continuous spectra is much greater than the fractional range in distance $\Delta R/R$ due to the annual variation in earth-sun distance. However, oscillations for monoenergetic solar neutrinos would produce observable annual variations in $\phi(R)$. In the standard solar model,[2] about 80% of the expected counting rate in the ^{37}Cl experiment[6] arises from ^{8}B neutrinos having a continuous spectrum extending from 0 to 14 MeV. The monoenergetic neutrinos making the greatest contribution to the expected counting rate are the 0.86-MeV neutrinos from ^{7}Be decay, which are expected to contribute 0.99 SNU.[6] Hence, if neutrino oscillations are occurring, one might expect to find a neutrino counting rate that varies with R according to

$$N(R) = \alpha_1 + \alpha_2 \cos kR, \tag{2}$$

where $\alpha_1 \approx \tfrac{1}{2}(4.7) \approx 2.4$ SNU and $|\alpha_2| \lesssim \tfrac{1}{2}(.99) \approx 0.50$ SNU (see Fig. 3). The maximum expected value for $|\alpha_2| \approx 0.50$ SNU would arise if the neutrino oscillation wavelength were less than the seasonal earth-sun distance variation, i.e., $\dfrac{\Delta R}{\lambda} > 2\pi$. In general, the expected amplitude of the cosine term, α_2, also depends on the phase of the oscillations when the earth is midway between perihelion and aphelion, i.e., $R_0 = \tfrac{1}{2}(R_A + R_P)$.

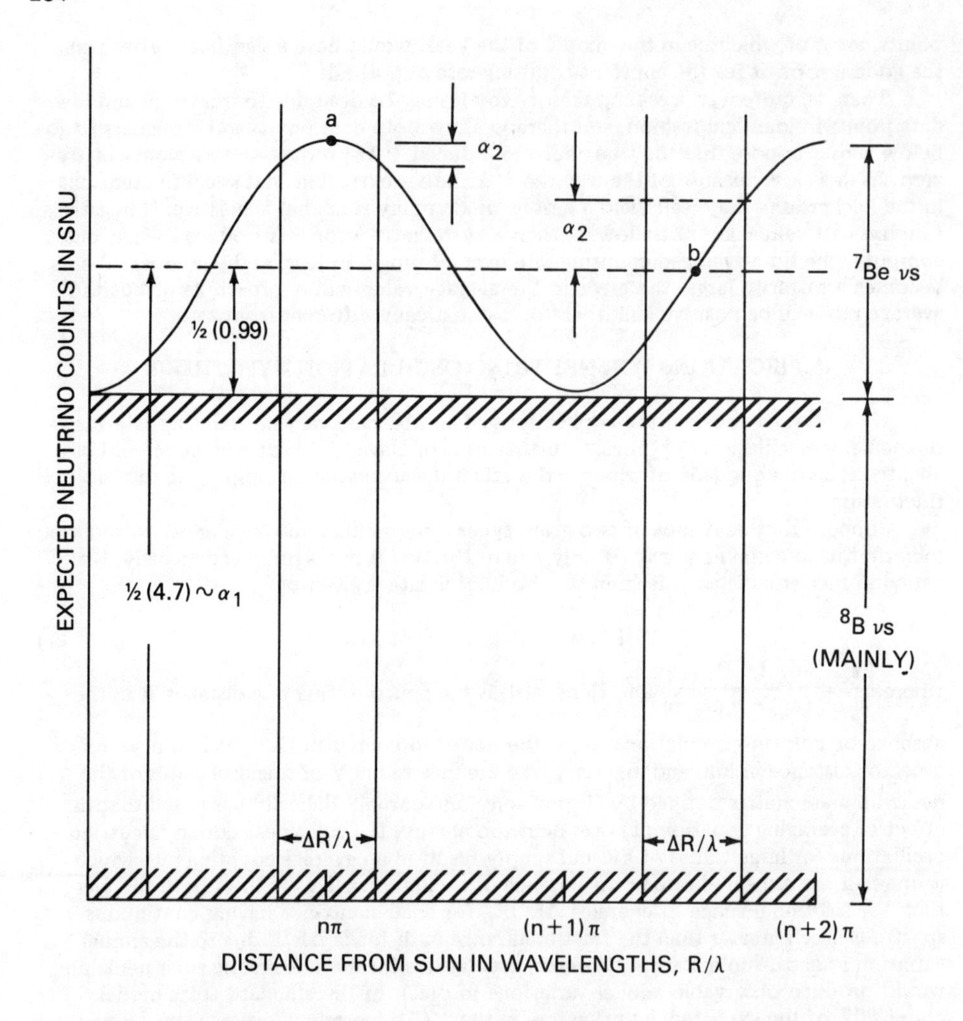

Fig. 3. Expected neutrino counting rate in SNU as a function of distance from the sun in neutrino oscillation wavelengths, R/λ. The continuous spectrum neutrinos (mainly from ^8B) give a counting rate independent of R, due to averaging Eq. (1) over neutrino energy. The monochromatic neutrinos (mainly from ^7Be) show an oscillation of amplitude α_2, which depends on both the ratio $\Delta R/\lambda = (R_A - R_p)/\lambda$, and on the phase of neutrino oscillations when the earth is midway between perihelion and aphelion, $R_o = \frac{1}{2}(R_p + R_A)$. For $\Delta R/\lambda < 1$, the maximum value of the oscillation amplitude, α_2, occurs for $kR_o = (n + \frac{1}{2})\pi$ (R_o/λ at point b), and the minimum value of the oscillation amplitude occurs for $kR_o = n\pi$ (R_o/λ at point a).

As indicated in Figs. 3 and 4, the largest expected amplitude occurs if $kR_o = (n+½)\pi$, and the smallest expected amplitude α_2 occurs if $kR_o = n\pi$. For neutrino oscillation wavelengths less than ΔR, the expected value of α_2 is 0.50 SNU, independent of phase. However, due to the fact that oscillations are averaged over each run, about half of which last over 100 days, a drop in observable amplitude occurs (indicated by the dotted curve in Fig. 4) for oscillation wavelengths much below ΔR.

In Fig. 5, Davis' data, corrected for varying solid angle subtended at the detector, are shown as a function of R, the earth-sun distance in AU. For each datum point, a value of R is obtained by averaging over the time interval of the run. The black data points show individual runs and the circles show averages over all runs occurring in four equal intervals in R. Combining groups of runs gives smaller vertical error bars and, moreover, conforms to the fact that many runs cover a significant fraction of a year. Three fits to the circled points are shown in Fig. 5. The poorest of these fits, with a χ^2 probability of 10%, is that to a nonvarying counting rate of 1.5 SNU. While this probability is not so low as to reasonably rule out the hypothesis of a constant counting rate, better fits do result to functions having the form of Eq. (2). Two such fits are shown in Fig. 5 for neutrino oscillation wavelengths equal to ΔR (curve a) and $2\Delta R$ (curve b). Clearly, the data are not sufficiently precise to give much information on a value for the neutrino wavelength, when even the presence of neutrino oscillations is open to question. It should, however, again be noted that the statistical evidence becomes considerably stronger if the errors on the low data points (and, in particular, the last circled point in Fig. 5) are significantly reduced, on the grounds discussed earlier. Moreover, some additional information on the neutrino oscillation wavelength results from the requirement that the maximum value for the amplitude should be ~0.50 SNU. This would favor curve (a) ($\alpha_2 = 0.98 \pm .55$ SNU) over curve (b) ($\alpha_2 = 2.4 \pm 1.2$ SNU), indicating that the oscillation wavelength is in the neighborhood of ΔR.

The expected value of the constant term in Eq. (2), $\alpha_1 = 2.4$ SNU is somewhat high, when compared to the average of the four circled data points, 1.5 SNU. This difference may be statistical or it may be narrowed by variations on the standard solar model. Alternatively, the discrepancy can also be eliminated through any mechanism that reduces the expected counting rate of the ^8B neutrinos somewhat, while not substantially affecting the expected counting rate for the ^7Be neutrinos. Leiter and Glass[9] have, in fact, suggested just such a mechanism based on a postulated nonminimal gravitational coupling (NMGC) for finite mass fermions. In another recent article, Mann and Primakoff[10] consider the form of neutrino oscillations, given the existence of an arbitrary number of neutrino types. Under certain conditions, the time-averaged neutrino flux is reduced by $1/n_\nu$, where n_ν is the number of "communicating" neutrino types. Thus, the expected counting rate can be reduced to any desired value for the ^8B neutrinos by postulating a sufficiently large number of neutrino types. Unfortunately, however, any oscillations produced by the ^7Be neutrinos would then probably be unobservable in Davis' data. For example, if there exist three types of neutrinos, then for the expected value of α_1 in equation (2), we find α_1 may be reduced to $(1/3)4.7 \approx$ 1.6 SNU. For the mononergetic ^7Be neutrinos, the equation analogous to Eq. (1) now has a linear combination of three cosine terms, corresponding to all possible pairs of the three neutrino types.[10] The maximum amplitude of one cosine term is given[10] by $(2/9)\phi(0) = 0.22 \times 0.20 \times 4.7 = 0.21$ SNU, a value too low to produce observable oscillations in Davis' data.

We summarize our results as follows:

1. The data from Davis' ^{37}Cl solar neutrino experiment is interpreted here as giving marginal evidence favoring the hypothesis of neutrino oscillations.

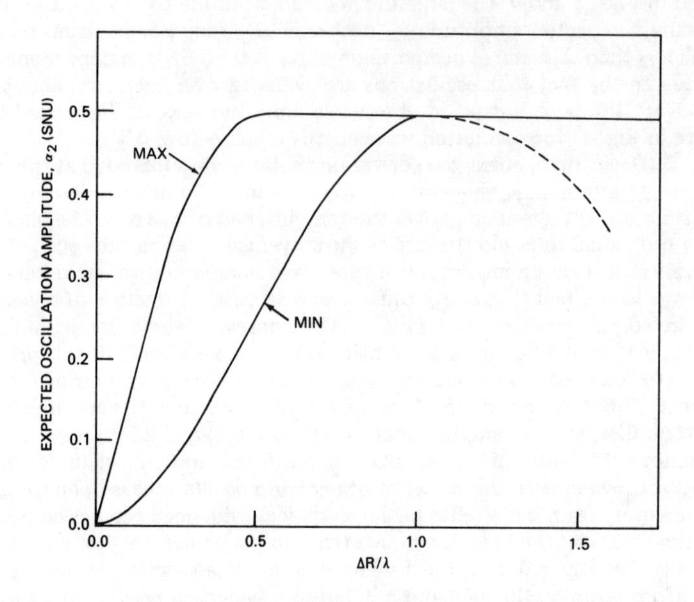

Fig. 4. Minimum and maximum expected amplitude α_2 for the cosine term in Eq. (2), as a function of the ratio $\Delta r/\lambda$. Here, λ is the neutrino oscillation wavelength and Δr is the earth-sun distance variation between perihelion and aphelion. The fall-off for $\Delta r/\lambda \gtrsim 1$ (dotted curve) is due to neutrino oscillations being averaged over during some of the long runs in Davis' experiment, and it only indicates a trend, not the actual numerical prediction.

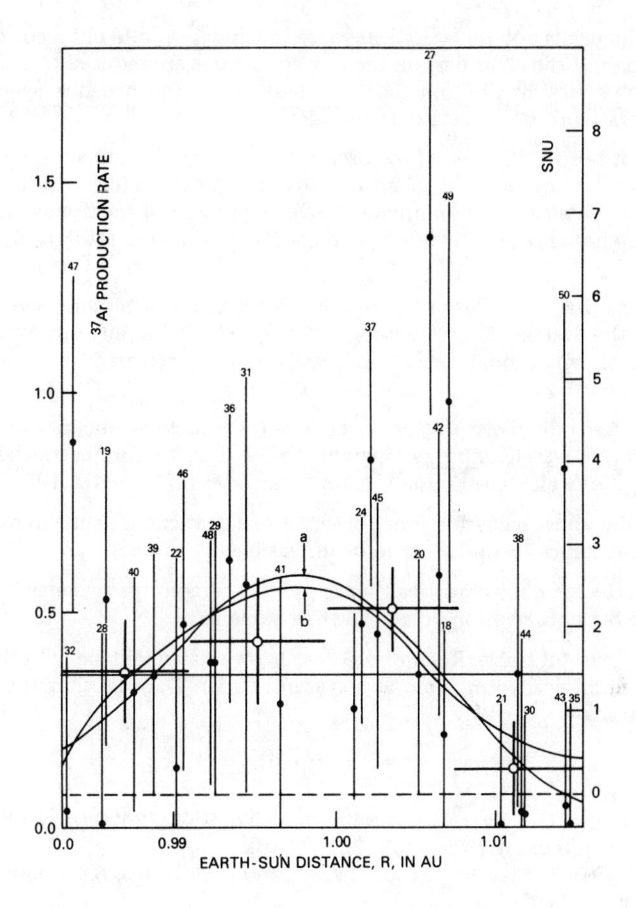

Fig. 5. Davis' data, corrected for varying solid angle subtended at the detector during the course of a year and plotted on a single annual scale in terms of earth-sun distance, R (in AU), instead of time. The R value for each run has been averaged over the time interval for that run. The white (circled) points show the average neutrino counting rate in four equal intervals of R. The horizontal line is a fit to the circled points assuming a constant counting rate. Curves (a) and (b) are fits assuming a variation of the form of Eq. (2), with neutrino oscillation wavelengths equal to ΔR (curve a) and 2ΔR (curve b), respectively.

2. The evidence is not yet statistically strong enough to rule out a constant counting rate, although, if the error bars on the low count runs are reduced to include only 34% of the probability distribution above the best values, the evidence against a constant counting rate probably is statistically significant.

3. A test of the significance of the oscillations can be made in a fairly short time with the present ^{37}Cl experiment. If future runs are kept short (close to the ^{37}Ar half life), much more data can be accumulated in a given period of time, since the error bars on long runs are not significantly less than on short runs, due to saturation effects for the ^{37}Ar.

4. Concerns that Davis has previously expressed about uncertainties in backgrounds are relevant to the average rate he reports (1.6 ± 0.4 SNU), but any observation of a statistically significant annual oscillation is unlikely to be affected by background uncertainties.

5. If the effect discussed in this paper is not a statistical fluctuation, then the neutrino oscillation wavelength is comparable to 5×10^6 km for neutrinos having an energy of 0.87 MeV, yielding a value for $\Delta m^2 = m_{\nu_1}{}^2 - m_{\nu_2}{}^2 \sim 4 \times 10^{-10}$ eV.^{2}

6. Given the above value for Δm^2, observations of vacuum neutrino oscillations over terrestrial distances would seem to be unfeasible.

7. If the effect is not a statistical fluctuation, it constitutes evidence that electron and muon number conservation breaks down at some level.

I am very grateful to Dr. Raymond Davis, Jr., who supplied me with the data for runs 18-50 in numerical form. I am also grateful to Dr. Darryl Leiter for his many helpful discussions.

FOOTNOTES AND REFERENCES

1. R. Davis, Jr. and J. C. Evans, Jr., BNL Report No. BNL 22920 (unpublished).
2. B. Kuchowicz, Rep. Prog. Phys. 39, 291 (1976).
3. J. K. Rowley, B. T. Cleveland, R. Davis, Jr., and J. C. Evans, BNL Report No. BNL 23418 (unpublished).
4. J. N. Bahcall, Astrophys. J. 216, L115 (1977).
5. V. Gribov and B. Pontecorvo, Phys. Lett. 28B, 493 (1969).
6. Figure 1 was taken from Ref. 3. The numerical values corresponding to these data plus three additional runs 48-50 were kindly supplied by Dr. Raymond Davis, Jr.
7. According to Dr. Davis, (private communication) the average value he reports does not contain such a systematic error, since it is obtained not by averaging data points for all runs, but by combining the raw data for all runs and doing a maximum likelihood fit to a constant counting rate plus an exponential decay.
8. J. N. Bahcall and S. C. Frautschi, Phys. Lett. 29B, 623 (1969).
9. D. Leiter and E. N. Glass, Phys. Rev. D 16, 3380 (1977). See also D. Leiter, these proceedings.
10. A. K. Mann and H. Primakoff, Phys. Rev. D 15, 655 (1977).

FERMION NONMINIMAL GRAVITATIONAL COUPLING AND THE SOLAR-NEUTRINO PROBLEM*

Darryl Leiter
Physics Department, George Mason University, Fairfax, Virginia 22030

ABSTRACT

In this work, a universal magnetic-dipole interaction between massive fermions is considered, with the coupling mediated by the local spacetime curvature. The strong principle of equivalence is not valid for fermions because of their intrinsic spin. Hence, the associated principle of "minimal gravitational coupling" for the Dirac equation coupled to electromagnetic fields in the presence of gravity is an assumption which is unsupported by either theory or experiment. We show that relaxing the arbitrary minimal-coupling constraint leads to a simple kind of nonminimal gravitational coupling (NMGC) which can generate curvature-dependent magnetic-moment effects, in background gravitational fields, for fermions coupled electromagnetically. Application of this model to the case of solar neutrinos yields a simple explanation of the low terrestrial neutrino flux in terms of sufficient neutrino energy loss (via multiple neutrino-electron magnetic elastic scattering in the solar plasma) to account for the very low detection rate on earth. Terrestrial tests of this type of NMGC effect for neutrinos would appear in high-energy neutrino-nucleon scattering, in terms of anomalous (charge-dependent) neutrino-deuteron interactions, which could not be explained by charge-independent neutral-current models alone.

INTRODUCTION

The strong principle of equivalence (SPE) is not valid for fermions because of their intrinsic spin. Hence, the associated principle of "minimal gravitational coupling" for the Dirac-Maxwell-Einstein system of coupled equations is an arbitrary assumption (usually justified on the basis of simplicity) which is unsupported by either theoretical foundations or experimental facts. <u>We show that, by eliminating this arbitrary assumption for fermions, we can construct a very simple form of "nonminimal gravitational coupling" (NMGC) which can generate curvature dependent magnetic moment effects for fermions in the presence of gravitational and electromagnetic fields. This model has implications for Davis' well known "solar neutrino problem." Application of this theory to the case of solar neutrinos yields a simple explanation for the extremely low terrestrial counting rate from the sun, in terms of a sufficient neutrino energy loss, due to multiple neutrino-electron magnetic elastic scattering in the solar plasma.</u>

These results can be obtained within the limits of the presently observed upper limits for neutrino-electron magnetic elastic scattering in terrestrial laboratories. Further terrestrial tests of this type of NMGC for neutrinos could appear in high energy neutrino-nucleon scattering processes, in terms of anomalous charge dependences in neutrino-deuteron scattering, which could not be explained by neutral-current models alone. This form of NMGC, assumed to be universal and extremely weak, would manifest itself most obviously for neutrinos, while being masked for particles carrying electromagnetic or hadronic charges. The philosophy of this approach, that one should never impose constraints (in this case the minimal-gravitational-coupling assumption for fermions) on theoretical models which are not experimentally verified, is reminiscent of

*See also D. Leiter and E. N. Glass, Phys. Rev. D <u>16</u>, 3380 (1977).

ISSN:0094-243X/79/520139-05$1.50

the situation regarding the assumption about parity conservation for neutrinos in weak interactions. Hence, the fact that NMGC is not ruled out theoretically for neutrinos, requires that it be ultimately up to experiment to determine the existence or nonexistence of NMGC. Our theoretical study is an attempt to induce such an experimental investigation.

A SIMPLE FORM OF NONMINIMAL GRAVITATIONAL COUPLING IN THE DIRAC-EINSTEIN-MAXWELL SYSTEM

If we eliminate the arbitrary assumption of minimal gravitational coupling in the Dirac-Einstein-Maxwell equations, consistent with the fact that the strong principle of equivalence[1] (SPE) is not valid for fermions, the simplest gauge-invariant[2] nonminimal gravitational interaction which mixes all three fields (and has dimensions of an energy density) is a magnetic-moment interaction which is curvature dependent in the form

$$\mathcal{L}_{NMGC} = \eta R(\tfrac{1}{2}\mu_0)\overline{\Psi}\sigma_{\mu\nu}\Psi F^{\mu\nu}, \tag{1}$$

where η has dimensions of $(\text{length})^2$ and $\mu_0 \equiv e/2m_e$. Units of magnetic moment are used and $\hbar = c = 1$. Inserting (1) into the standard Dirac-Einstein-Maxwell action yields

$$I = \int d^4x(-g)^{1/2}\,[R + \kappa(\mathcal{L}_{Dirac} + \tfrac{1}{4}F_{\mu\nu}F^{\mu\nu} + J_\mu A^\mu + \mathcal{L}_{NMGC})], \tag{2}$$

where the NMGC term has a strength determined by the value of $\eta\kappa$, and is added directly into the original minimally coupled action principle. Then variation $\delta I = 0$ with respect to $\delta g_{\mu\nu}$, δA_μ, and $\delta\Psi$ gives,[3] respectively, a modified energy-momentum tensor in the Einstein equation

$$G_{\mu\nu} = -\kappa T_{\mu\nu}(\eta),$$

$$T_{\mu\nu} = (1 + \eta\kappa S)^{-1}\left[t_{\mu\nu(\text{Dirac})} + t_{\mu\nu(\text{Maxwell})} + \eta\left(R\frac{\partial S}{\partial g^{\mu\nu}} + g_{\mu\nu}S_{;\alpha}{}^{\alpha} - S_{;\mu\nu}\right)\right], \tag{3}$$

where $S = \frac{1}{2}\mu_0\overline{\Psi}\sigma_{\mu\nu}\Psi F^{\mu\nu}$ and the Einstein gravitational constant is $\kappa = 8\pi Gc^{-4}$; a modified current in the Maxwell equation

$$F_{\mu\nu}{}^{;\nu} = J_\mu(\eta) = [e\overline{\Psi}\gamma_\mu\Psi - (\tfrac{1}{2}\eta R\mu_0\overline{\Psi}\sigma_{\mu\nu}\Psi)^{;\nu}]; \tag{4}$$

and, finally, a modified Dirac equation with a curvature-dependent anomalous magnetic moment,

$$(-i\gamma^\mu\nabla_\mu + m + e\not{A} + \tfrac{1}{2}\eta R\mu_0\sigma_{\mu\nu}F^{\mu\nu})\Psi = 0. \tag{5}$$

Note that the NMGC terms do not vanish in a local geodesic frame, as expected.

APPLICATION TO THE SOLAR NEUTRINO PROBLEM

From the preceding section we see that the effect of the NMGC term (1) on the action (2) is to give the fermion an anomalous magnetic moment $\mu = \eta R\mu_0$, depending on the scalar curvature R.[4] Let us suppose that we choose the value of η subject to the constraint that this effect is dwarfed by the presence of the much stronger electromagnetic and strong interactions (which would account for our lack of any experimental evidence of this term for charged leptons and hadrons). Then it still might be observable for neutrinos, since they have zero charge, and only a very weak interaction with matter.

To test this hypothesis, we will apply it to the case of the solar neutrinos. The possibility of the neutrinos undergoing multiple magnetic elastic scattering on the electrons in the solar plasma has been previously investigated[5] as a possible solution to the "solar neutrino problem." Since there exists some evidence that neutrinos may have a small nonzero mass,[6] we will assume in our calculations that the neutrino is described by (5) with e = 0 and $m_\nu < 60$ eV. We take the general point of view[7] that parity nonconservation is a property of the weak-interaction matrix element itself [hence, if Ψ is a solution to (5) with e = 0 and $m_\nu \neq 0$, then only $\frac{1}{2}(I - i\gamma^5)\Psi$ contributes to the weak-interaction matrix element (even if $m_\nu \neq 0$)]. In the sun, the metric is slowly varying over microscopic dimensions, so we can calculate effectively in a local freely falling geodesic frame where the first derivatives of $g_{\mu\nu}$ vanish. If we also assume that gravity is treated as an external field, then the equations for solar neutrinos inside the solar electron-proton plasma can be specialized from (5) to the simpler form (6) below under the following assumptions. First, for solar neutrinos $E_\nu < 10$ MeV, so this means that $\lambda_\nu \sim \hbar c/E_\nu > 10^{-12}$ cm; thus, the protons are seen by the neutrinos as if they were massive charged pointlike structures surrounded by unbound negatively-charged electron distributions. Hence, the proton dynamics can be neglected at this energy and only neutrino-electron scattering will be important.[8] Second, the universal NMGC is assumed to be masked, in the electron equation, by the electromagnetic term $-e\not{A}$. Hence, it is neglected in this approximation. Under these assumptions, the local equations for NMGC solar neutrinos in the solar plasma are approximated to first order in $\eta\kappa$ as

$$(-i\not{\partial} + m_{(\nu)} + \tfrac{1}{2}\eta R\mu_0\sigma_{\mu\nu}F^{\mu\nu})\Psi_{(\nu)} = 0, \tag{6a}$$

$$(-i\not{\partial} + m_e - e\not{A})\Psi_e = 0, \tag{6b}$$

$$F_{\mu\nu}{}^{,\nu} = -e\Psi_e\gamma_\mu\Psi_e - \tfrac{1}{2}(\eta R\mu_0\overline{\Psi}_{(\nu)}\sigma_{\mu\nu}\Psi_{(\nu)})^{,\nu}, \tag{6c}$$

$$R \cong \kappa(T_e + T_p)$$

$$\cong \kappa c^2(S_e + S_p), \tag{6d}$$

where $\mu_0 \equiv e^2/2m_e$ and S_e, S_p are the electron and proton mass densities.

From (6a), we see that for the solar neutrinos, specifically, the scalar curvature in (6d) generates an anomalous magnetic moment term, so that the neutrino equation of motion is

$$[-i\not{\partial} + m_{(\nu)} + \tfrac{1}{2}\eta\kappa c^2\mu_0(S_e + S_p)\sigma_{\mu\nu}F^{\mu\nu}]\Psi_{(\nu)} = 0. \tag{7}$$

However, in the neutral solar plasma, since $\lambda_\nu > 10^{-12}$ cm, S_p is seen by the neutrino to be a superposition of pointlike proton mass densities surrounded by spread out $S_e(x)$ electron-wave mass densities. Since the average density of the solar plasma ~ 1 g/cm^3, this implies that there are 10^{25} protons/cm^3. Hence, there is one proton every 10^{-8} cm along the neutrino path. The average neutrality of the plasma then implies that there is also one electron wave spread out over the 10^{-8} cm in between each pointlike proton, as seen by the solar neutrinos moving through the plasma. Thus, in (7) only the spread out structure of the electron clouds will be sensed most sensitively by neutrinos with $\lambda_\nu > 10^{-12}$ cm (while the more massive point-like protons will exert a much smaller effect). For elastic scattering, most of the recoil from neutrino scattering will be taken up by the electrons[8] and for this case (7) can be written in the more revealing form

$$[-i\not{\partial} + m_{(\nu)} + \tfrac{1}{2}\mu_{(\nu)}(x)\sigma_{\mu\nu}F^{\mu\nu}]\Psi_{(\nu)} = 0, \tag{8}$$

where $\mu_{(\nu)}(x)$ is the effective NMGC-induced neutrino magnetic moment inside the solar plasma which participates in neutrino-electron magnetic elastic scattering:

$$\mu_{(\nu)}(x) \simeq [\eta\kappa c^2 S_e(x)]\mu_0. \tag{9}$$

Now, $|S_e| \simeq [10^{-28}/(10^{-8})^3] \simeq 10^{-4}$ g/cm^3 in the solar plasma, so that the NMGC-induced $\mu_{(\nu)}(x)$ is spread out over 10^{-8} cm with an average density of 10^{-4} g/cm^3. Hence, we can write $\mu_{(\nu)}(x)$ in the form

$$\mu_{(\nu)}(x) \simeq (\eta\kappa c^2\mu_0) \times 10^{-4} F(x), \qquad \int F(x)dx^3 = 1, \tag{10}$$

where the density structure F(x) acts like a "magnetic form factor."

Applying (8) through (10) to the work of Clark and Pedigo,[5] we see that the NMGC effect gives the neutrino a very large spread for its effective magnetic form factor $\sim 10^{-8}$ cm. This is precisely the kind of effect which enables the neutrino-electron multiple magnetic-elastic scattering models to account for sufficient neutrino energy loss to fall below the detection threshold of the Davis experiment.[9] To see this in more detail, we note that the NMGC neutrino equations (8), (9), and (10) can be written as

$$[-i\partial\!\!\!/ + m_\nu + \eta\kappa c^2\mu_0 S_e \sigma_{\mu\nu}\partial^\nu A^\mu]\Psi_{(\nu)} = 0. \tag{11}$$

To lowest order, the magnetic-elastic scattering matrix element for the neutrinos and electrons is of the same form as that used by Clark and Pedigo, except for the fact that our magnetic form factor F(x) is related to the local electron mass density in the solar plasma as

$$S_e(x) \simeq m_e\overline{\Psi}_e\Psi_e \simeq 10^{-4}F(x). \tag{12}$$

Hence, we may directly use their approach to offer a possible solution to the solar neutrino problem. Since they use the results of Cowan and Reines,[10] that experimentally for terrestrial $\nu_e - e$ scattering $\mu_{(\nu)} < 10^{-9}\mu_0$, this means that from (10) we can determine an upper limit on $\eta\kappa$ as

$$(\eta\kappa)c^2 10^{-4} < 10^{-9},$$

which yields an upper limit of

$$\eta\kappa < 10^{-26}. \tag{13}$$

This is the terrestrial upper limit on the coupling $\eta\kappa$ of the NMGC-induced magnetic moment of the neutrino. Since our model predicts a spread out "solar plasma" form factor $\langle r\rangle_\nu \sim 10^{-8}$ cm, then the results of Clark and Pedigo[5] will satisfactorily hold and indicate that choices of F(x) which imply $\langle r\rangle_{(\nu)} \gg 7 \times 10^{-10}$ cm are explainable by the NMGC hypothesis.

Recently, Davis and Evans[11] have reported a more reliable solar neutrino counting rate of 1.6 ± 0.4 SNU (1 SNU = 10^{-36} captures per atom per second). This deviation from the standard solar model might be accounted for by any mechanism which reduces the expected counting rate of the ^{8}B neutrinos, since 80% of the standard counting rate comes from this source (a continuous spectrum extending from 0 to 14 MeV). Since the NMGC neutrino magnetic moment in (10) is density dependent (via $S_e(x)$) it is plausible to expect that the ^8B neutrinos may be most severely affected by the NMGC magnetic-elastic scattering effect, since they are generated deep within the solar core, where the densities are at a maximum. Hence, the possibility arises that NMGC may play a significant role in future explanations of the recent Davis data, since it may be a simple and elegant way to explain the results in terms of a significantly reduced ^8B neutrino signal at the earth.

FOOTNOTES AND REFERENCES

1. Our statement of the SPE is that at each point of space-time the gravitational field variables can be removed from the field equations of matter by a coordinate transformation. This is equivalent to demanding that there be no explicit appearance of curvature terms in the interaction Lagrangian.
2. By "simplest," we mean that this term is linear in R and $F^{\mu\nu}$, and bilinear in Ψ, which is consistent with gauge invariance of both the first and second kinds. It is interesting to note that one might think that a term such as $\eta R\bar{\Psi}\gamma_\mu \Psi A^\mu$ would be simpler; however, this term is not gauge invariant because of $R_{,\mu} \neq 0$ in the general context.
3. Details on the explicit formalism associated with the generally covariant Dirac equation and variational principles which involve the Dirac-Einstein system are given in D. Leiter, Lett. Nuovo Cimento 12, 633 (1975); D. Leiter and T. Chapman, Am. J. Phys. 44, 858 (1976). See H. C. Ohanian, J. Math. Phys. 14, 1892 (1973), for comments on the relationship of the SPE to general concepts of NMGC.
4. The scalar curvature R is obtained from Eq. (3) by taking the trace of the Einstein equation over the space-time indices.
5. R. B. Clark and R. D. Pedigo, Phys. Rev. D 8, 2261 (1973).
6. V. E. Barnes et al., Phys. Rev. Lett. 38, 1049 (1977).
7. This is the well-known idea that parity violation in weak interactions may be connected with the inherent chirality invariance of the associated Lagrangian, and not necessarily to the requirement of zero neutrino mass.
8. This is the same approximation as that used by Clark and Pedigo in Ref. 5.
9. Good general discussions of this are given in the review article by B. Kuchowicz, Rep. Prog. Phys. 39, 291 (1976), and in the lectures of R. T. Rood, proceedings of the 1976 Summer School "Ettore Majorana," Erice, Italy (unpublished). See also BNL Report No. BNL-21837 (unpublished) for the latest analysis of the data by R. Davis, Jr., and his group.
10. C. L. Cowan and F. Reines, Phys. Rev. 107, 528 (1957).
11. R. Davis, Jr., and C. J. Evans, Jr., BNL Report No. BNL 22920, 1977, (unpublished).

AIP Conference Proceedings

No.	Title	L.C. Number	ISBN
No.1	Feedback and Dynamic Control of Plasmas (Princeton) 1970	70-141596	0-88318-100-2
No.2	Particles and Fields - 1971 (Rochester)	71-184662	0-88318-101-0
No.3	Thermal Expansion - 1971 (Corning)	72-76970	0-88318-102-9
No.4	Superconductivity in d- and f-Band Metals (Rochester, 1971)	74-18879	0-88318-103-7
No.5	Magnetism and Magnetic Materials - 1971 (2 parts) (Chicago)	59-2468	0-88318-104-5
No.6	Particle Physics (Irvine, 1971)	72-81239	0-88318-105-3
No.7	Exploring the History of Nuclear Physics (Brookline, 1967, 1969)	72-81883	0-88318-106-1
No.8	Experimental Meson Spectroscopy - 1972 (Philadelphia)	72-88226	0-88318-107-X
No.9	Cyclotrons - 1972 (Vancouver)	72-92798	0-88318-108-8
No.10	Magnetism and Magnetic Materials - 1972 (2 parts) (Denver)	72-623469	0-88318-109-6
No.11	Transport Phenomena - 1973 (Brown University Conference)	73-80682	0-88318-110-X
No.12	Experiments on High Energy Particle Collisions - 1973 (Vanderbilt Conference)	73-81705	0-88318-111-8
No.13	π-π Scattering - 1973 (Tallahassee Conference)	73-81704	0-88318-112-6
No.14	Particles and Fields - 1973 (APS/DPF Berkeley)	73-91923	0-88318-113-4
No.15	High Energy Collisions - 1973 (Stony Brook)	73-92324	0-88318-114-2
No.16	Causality and Physical Theories (Wayne State University, 1973)	73-93420	0-88318-115-0
No.17	Thermal Expansion - 1973 (Lake of the Ozarks)	73-94415	0-88318-116-9
No.18	Magnetism and Magnetic Materials - 1973 (2 parts) (Boston)	59-2468	0-88318-117-7
No.19	Physics and the Energy Problem - 1974 (APS Chicago)	73-94416	0-88318-118-5
No.20	Tetrahedrally Bonded Amorphous Semiconductors (Yorktown Heights, 1974)	74-80145	0-88318-119-3
No.21	Experimental Meson Spectroscopy - 1974 (Boston)	74-82628	0-88318-120-7
No.22	Neutrinos - 1974 (Philadelphia)	74-82413	0-88318-121-5
No.23	Particles and Fields - 1974 (APS/DPF Williamsburg)	74-27575	0-88318-122-3
No.24	Magnetism and Magnetic Materials - 1974 (20th Annual Conference, San Francisco)	75-2647	0-88318-123-1
No.25	Efficient Use of Energy (The APS Studies on the Technical Aspects of the More Efficient Use of Energy)	75-18227	0-88318-124-X
No.26	High-Energy Physics and Nuclear Structure - 1975 (Santa Fe and Los Alamos)	75-26411	0-88318-125-8

No.27	Topics in Statistical Mechanics and Biophysics: A Memorial to Julius L. Jackson (Wayne State University, 1975)	75-36309	0-88318-126-6
No.28	Physics and Our World: A Symposium in Honor of Victor F. Weisskopf (M.I.T., 1974)	76-7207	0-88318-127-4
No.29	Magnetism and Magnetic Materials - 1975 (21st Annual Conference, Philadelphia)	76-10931	0-88318-128-2
No.30	Particle Searches and Discoveries - 1976 (Vanderbilt Conference)	76-19949	0-88318-129-0
No.31	Structure and Excitations of Amorphous Solids (Williamsburg, Va., 1976)	76-22279	0-88318-130-4
No.32	Materials Technology - 1975 (APS New York Meeting)	76-27967	0-88318-131-2
No.33	Meson-Nuclear Physics - 1976 (Carnegie-Mellon Conference)	76-26811	0-88318-132-0
No.34	Magnetism and Magnetic Materials - 1976 (Joint MMM-Intermag Conference, Pittsburgh)	76-47106	0-88318-133-9
No.35	High Energy Physics with Polarized Beams and Targets (Argonne, 1976)	76-50181	0-88318-134-7
No.36	Momentum Wave Functions - 1976 (Indiana University)	77-82145	0-88318-135-5
No.37	Weak Interaction Physics - 1977 (Indiana University)	77-83344	0-88318-136-3
No.38	Workshop on New Directions in Mössbauer Spectroscopy (Argonne, 1977)	77-90635	0-88318-137-1
No.39	Physics Careers, Employment and Education (Penn State, 1977)	77-94053	0-88318-138-X
No.40	Electrical Transport and Optical Properties of Inhomogeneous Media (Ohio State University, 1977)	78-54319	0-88318-139-8
No.41	Nucleon-Nucleon Interactions - 1977 (Vancouver)	78-54249	0-88318-140-1
No.42	Higher Energy Polarized Proton Beams (Ann Arbor, 1977)	78-55682	0-88318-141-X
No.43	Particles and Fields - 1977 (APS/DPF, Argonne)	78-55683	0-88318-142-8
No.44	Future Trends in Superconductive Electronics (Charlottesville, 1978)	77-9240	0-88318-143-6
No.45	New Results in High Energy Physics - 1978 (Vanderbilt Conference)	78-67196	0-88318-144-4
No.46	Topics in Nonlinear Dynamics (La Jolla Institute)	78-057870	0-88318-145-2
No.47	Clustering Aspects of Nuclear Structure and Nuclear Reactions (Winnepeg, 1978)	78-64942	0-88318-146-0
No.48	Current Trends in the Theory of Fields (Tallahassee, 1978)	78-72948	0-88318-147-9
No.49	Cosmic Rays and Particle Physics - 1978 (Bartol Conference)	79-50489	0-88318-148-7
No.50	Laser-Solid Interactions and Laser Processing - 1978 (Boston)	79-51564	0-88318-149-5
No.51	High Energy Physics with Polarized Beams and Polarized Targets (Argonne, 1978)	79-64565	0-88318-150-9
No. 52	Long-Distance Neutrino Detection - 1978 (C.L. Cowan Memorial Symposium)	79-52078	0-88318-151-7